THE NIGHT SKY
MONTH BY MONTH

THE NIGHT SKY
MONTH BY MONTH

WILL GATER with GILES SPARROW

LONDON, NEW YORK, MELBOURNE, MUNICH, AND DELHI

DK UK

Editor	Martha Evatt
Designer	Duncan Turner
Managing Editor	Sarah Larter
Managing Art Editor	Michelle Baxter
Production Editor	Sophie Argyris
Production Controller	Phil Sergeant
Publishing Manager	Liz Wheeler
Art Director	Phil Ormerod
Reference Publisher	Jonathan Metcalf
Picture Researcher	Louise Thomas
Jacket Designer	Mark Cavanagh

DK INDIA

Editors	Soma B Chowdhury
	Sudeshna Dasgupta
	Himanshi Sharma
Designers	Nidhi Mehra
	Pooja Pipil
Project Editor	Alka Ranjan
DTP Designers	Vishal Bhatia
	Saurabh Challariya
	Pushpak Tyagi
Managing Editor	Rohan Sinha
Managing Art Editor	Ashita Murgai
DTP Manager	Sunil Sharma

This edition published in 2011
First published in 2011 by Dorling Kindersley Limited
80 Strand, London WC2R ORL
Penguin Group (UK)

Looking at the Sun with the naked eye, binoculars, or a telescope can cause eye damage. The authors and publishers cannot accept any liability for readers who do not take precautions to observe safely. Modifying cameras or other equipment may invalidate the manufacturers warranty and readers do so at their own risk.

Colour reproduction by MDP, Bath
Printed and bound in China by Hung Hing

Discover more at
www.dk.com

CONTENTS

LOOKING UP

MONTHLY SKY GUIDES

ALMANAC

CONTRIBUTORS

Will Gater is an astronomy journalist and author. He has written for several of the UK's top astronomy and science magazines and promotes both these subjects with frequent appearances on television and radio. His blog and website can be found at www.willgater.com. Will is also the author of *The Practical Astronomer*, published by Dorling Kindersley.

Giles Sparrow is a writer specializing in astronomy and space science. He has degrees in astronomy and science communication, and has worked for 15 years as an editor and author. An avid follower of the unfolding story of space exploration, he has written on space technology and the history of spaceflight in a wide range of books, including Dorling Kindersley's bestselling *Universe*.

LOOKING UP

● ●

By watching the night skies and observing stars, planets, nebulae, and galaxies, stargazers can start to understand the vast Universe and all it encompasses. Astronomers picture the night sky as a starry sphere around Earth – "the celestial sphere" – which allows them to find their way around it and track the movements of night-sky objects as the Earth rotates.

Leonid meteor shower
Shooting stars, here seen over Joshua Tree National Park in the USA, light up the sky. Eagerly anticipated every year, the Leonid meteor shower originates in the constellation Leo and annually peaks around 17 November.

LOOKING INTO SPACE

The Earth is a part of a much greater Universe than our eyes alone reveal to us. Knowing our place within it helps us to understand what we are seeing.

THE SCALE OF THE UNIVERSE

The Universe we live in is immense, extending far beyond the furthest edge of our Solar System. On a clear night, many stars are visible to the naked eye, and sometimes the faint glowing band of the Milky Way galaxy can be seen. From a very dark site, the Andromeda Galaxy, a staggering 2.5 million light-years away, is also visible. It is one of the most distant deep-sky objects visible to the naked eye. A telescope or a pair of binoculars allows us to see objects that are even further away, such as other galaxies, nebulae or star clusters in our galaxy, the Milky Way. However, there is much more to be seen, and amateur astronomers can only see a small fraction of it.

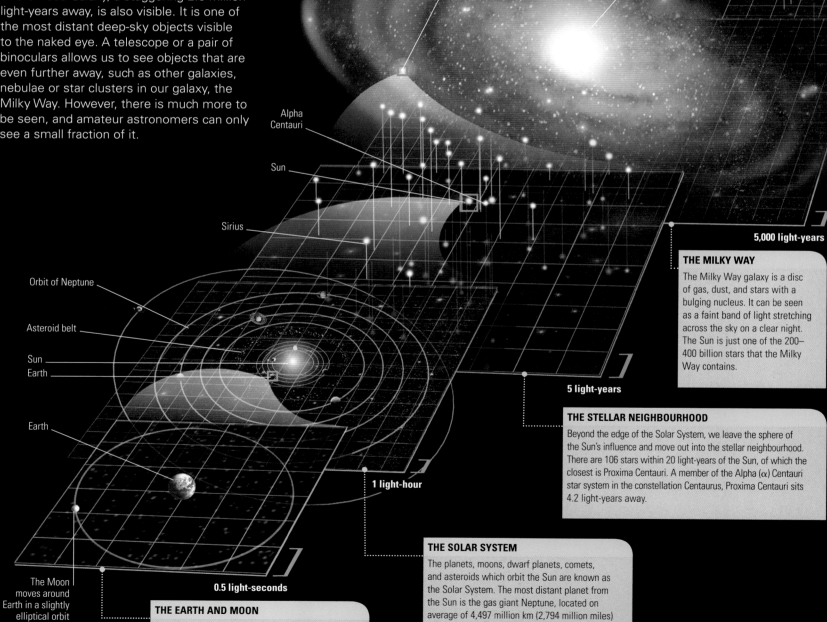

The Andromeda Galaxy lies 2.5 million light-years away from Earth, in the Milky Way

NGC 147
NGC 185
Andromeda I
Andromeda II
Andromeda III
Triangulum Galaxy

The stellar neighbourhood lies in the Orion Arm of the Milky Way, around 26,000 light-years from its centre

Galactic nucleus

Alpha Centauri

Sun

Sirius

5,000 light-years

Orbit of Neptune

Asteroid belt

Sun
Earth

Earth

1 light-hour

0.5 light-seconds

The Moon moves around Earth in a slightly elliptical orbit

THE MILKY WAY

The Milky Way galaxy is a disc of gas, dust, and stars with a bulging nucleus. It can be seen as a faint band of light stretching across the sky on a clear night. The Sun is just one of the 200–400 billion stars that the Milky Way contains.

5 light-years

THE STELLAR NEIGHBOURHOOD

Beyond the edge of the Solar System, we leave the sphere of the Sun's influence and move out into the stellar neighbourhood. There are 106 stars within 20 light-years of the Sun, of which the closest is Proxima Centauri. A member of the Alpha (α) Centauri star system in the constellation Centaurus, Proxima Centauri sits 4.2 light-years away.

THE SOLAR SYSTEM

The planets, moons, dwarf planets, comets, and asteroids which orbit the Sun are known as the Solar System. The most distant planet from the Sun is the gas giant Neptune, located on average of 4,497 million km (2,794 million miles) away from the Sun.

THE EARTH AND MOON

The Moon is the nearest celestial body to Earth, sitting 384,400km (238,900 miles) away. Light takes just over a second to reach Earth from the Moon.

THE LOCAL GROUP OF GALAXIES

The Milky Way is a part of a much larger gathering of around 40 galaxies that exist in the nearby Universe. These are known collectively as the Local Group. Some of the Local Group galaxies can be seen easily in the night sky, such as the Andromeda Galaxy, M31, and the Triangulum Galaxy, M33.

THE LOCAL SUPERCLUSTER

The Local Group is itself a part of a larger group, formed by thousands of galaxies. Known as the Virgo Supercluster, this swarm of galaxies is 100 million light-years wide. This supercluster nestles in vast interconnected filaments of other superclusters, which stretch across the Universe.

...a Minor dwarf galaxy

...e Milky Way

250,000 light-years

Leo A

10 million light-years

MEASURING DISTANCE

As the Universe is such a large place, the units of distance we use in everyday life, such as kilometres or miles, are not very helpful in communicating the vast distances between stars and galaxies. Instead, astronomers use units called light-years to mark the vast distances between the stars and galaxies. One light-year is equal to the distance that a ray of light travels over the course of one year. The speed of light is an incredible 300,000km (186,000 miles) per second, so one light-year is a huge distance. As the distances (see below) between the galaxies and even nearby stars are so great, light takes a long time to travel across space. When we see the light from an object like a star, it may have taken decades, centuries, or millions of years to reach us. Essentially, we are peering back in time, as we see the object as it was when that light left it, not what it looks like "now". So for an object like the Andromeda Galaxy, which is 2.5 million light-years away, we are seeing it as it was 2.5 million years ago. In contrast, the Sun's light takes just 8.5 minutes to reach earth. In the chart below, the first division represents 10,000km (6,200 miles). Each further division marks a 10x increase in scale.

Peering into the depths
The objects we see in the night sky are not all at the same distance from us. We can see everything from meteors shooting through our atmosphere to distant stars in our own galaxy.

DISTANCE FROM THE CENTRE OF EARTH

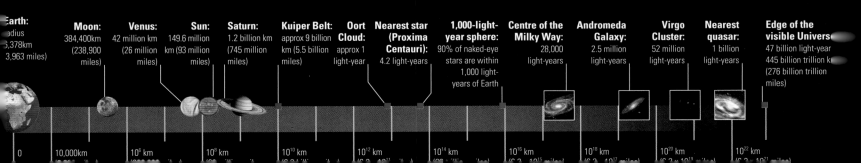

| Earth: radius 6,378km (3,963 miles) | Moon: 384,400km (238,900 miles) | Venus: 42 million km (26 million miles) | Sun: 149.6 million km (93 million miles) | Saturn: 1.2 billion km (745 million miles) | Kuiper Belt: approx 9 billion km (5.5 billion miles) | Oort Cloud: approx 1 light-year | Nearest star (Proxima Centauri): 4.2 light-years | 1,000-light-year sphere: 90% of naked-eye stars are within 1,000 light-years of Earth | Centre of the Milky Way: 28,000 light-years | Andromeda Galaxy: 2.5 million light-years | Virgo Cluster: 52 million light-years | Nearest quasar: 1 billion light-years | Edge of the visible Universe: 47 billion light-years 445 billion trillion km (276 billion trillion miles) |

| 0 | 10,000km | 10⁶ km | 10⁹ km | 10¹⁰ km | 10¹² km | 10¹⁴ km | 10¹⁵ km | 10¹⁸ km | 10²⁰ km | 10²² km |

LOOKING AT STARS

When we look at the sky on a clear night we are greeted by countless stars. Stargazers use several methods to navigate their way around the magnificent night sky.

THE CELESTIAL SPHERE

The coordinate system of latitude and longitude helps us to locate an object on the Earth's surface. This system is based around a simple imaginary grid, where latitude – the horizontal lines on the grid – is measured north or south from the equator. Longitude, the vertical grid lines, is measured east or west from a point known as the prime meridian – a circle running through the North and South Poles and Greenwich in London. Similarly, astronomers project an imaginary sphere, also known as the celestial sphere, onto the sky. It has its own grid lines: the prime meridian, known as the "celestial meridian", and the equator, known as the "celestial equator". Instead of latitude, astronomers use declination, which is measured in degrees and minutes, while longitude becomes right ascension, or RA, and is measured in hours and minutes. These coordinates help astronomers to locate celestial objects in space.

OUR VIEW OF THE CELESTIAL SPHERE

Your view of the night sky depends on where you are on Earth's surface. If you are located in the southern hemisphere, you see a different portion of the celestial sphere to someone who is observing the night sky in the northern hemisphere. However, from the equator you can view the entire celestial sphere over the course of a year.

Limited view
Your location on Earth determines the part of the celestial sphere that you can see.

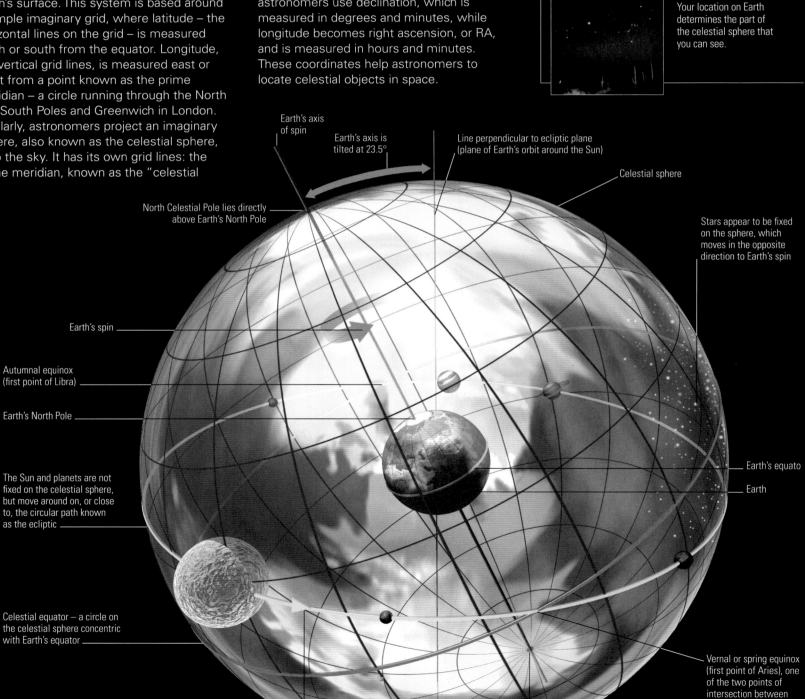

Earth's axis of spin

Earth's axis is tilted at 23.5°

Line perpendicular to ecliptic plane (plane of Earth's orbit around the Sun)

Celestial sphere

North Celestial Pole lies directly above Earth's North Pole

Stars appear to be fixed on the sphere, which moves in the opposite direction to Earth's spin

Earth's spin

Autumnal equinox (first point of Libra)

Earth's North Pole

The Sun and planets are not fixed on the celestial sphere, but move around on, or close to, the circular path known as the ecliptic

Earth's equator

Earth

Celestial equator – a circle on the celestial sphere concentric with Earth's equator

Vernal or spring equinox (first point of Aries), one of the two points of intersection between celestial and ecliptic

Defining positions

To create a coordinate system where objects can be located on the night sky, astronomers imagine that Earth is embedded in an imaginary

...EASURING SIZES

...hen observing the night sky for the first time ...an be tricky to gauge how big a constellation ...an object is by just looking at a star chart. ...wever, there are some easy ways to ...asure apparent sizes on the sky. Held at ...'s length against the background of the sky, ... hands and fingers can act as convenient ...asuring scales. For example, an index finger ...vers the Moon, which is only half a degree

across. While the width of an outstretched hand is roughly 22° across. Distances between objects on the night sky are measured in units called degrees. For example, the Andromeda Galaxy, or M31, appears roughly 3° across on the sky. A degree can be broken down into 60 arcminutes (with the symbol '), and each arcminute can be broken down into 60 arcseconds (with the symbol "). These units are sometimes written in slightly different ways and you might see them as minutes of arc, arcmin, seconds of arc, or arsec. These smaller units are often encountered when dealing with the separation between double stars or the size of a nebula or cluster.

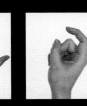

...dspan
...lly outstretched ...d held at arm's ...th spans about ... of the sky.

Finger joints
A side-on fingertip is about 3° wide; the second joint is 4°, the third joint 6°.

Finger width
One finger at arm's length will cover the Moon, which is less than 1° across.

- 1 degree
- 90 degrees
- 360 degrees

ANGULAR DISTANCES

OBJECT OR DISTANCE	APPROXIMATE ANGULAR SIZE
Distance from the pointers in the Plough to Polaris	28°
Distance between the Crux pointers	6°
Distance between the pointers in the Plough	5°
Your little finger at arm's length	1°
The Moon (average size)	31'
The Sun (average size)	32'
Distance between Jupiter and Ganymede (the brightest of its main moons)	6'
Resolution of the naked eye (this means the ability of your eye to split two objects that are as close together as this)	3' 25"

...ARHOPPING

... navigate around the night sky using a ...r chart, you can use a method known ... starhopping. This is a technique used by ...perts and beginners alike, and can come ...handy when you are trying to find faint ...escopic targets that may not be visible to ...e naked eye. The basic method is to first ...eck a star chart (see pp.16-114) that shows ...e object you are trying to find, as well as ...y bright stars nearby. Then, start by finding ...tar or pattern of stars in the sky that you

recognize and can easily locate. Once you find one recognizable star, you can then hop to another, possibly fainter star nearby, continuing to other stars until you eventually find your target. This is a great way to learn your way around the night sky, and is also handy when you use a pair of binoculars or a telescope. A more detailed printed chart from planetarium software will help you to find objects when you are starhopping using a telescope, such as faint galaxies.

UNDERSTANDING CELESTIAL COORDINATES

To understand right ascension and declination you need to know the reference points from which they are measured. The "zero" point for declination lies on a line on the celestial sphere called the celestial equator. You can think of it simply as Earth's equator projected onto the imaginary celestial sphere. Objects above the celestial equator, towards the North Celestial Pole, have a positive declination, and those below, towards the South Celestial Pole have negative declination. For right ascension, the zero mark is a line called the "celestial meridian" marked at the moment where the Sun crosses the celestial equator.

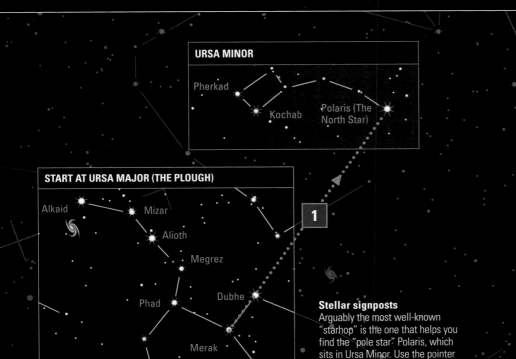

URSA MINOR

Pherkad

Kochab

Polaris (The North Star)

START AT URSA MAJOR (THE PLOUGH)

Alkaid · Mizar

Alioth

Megrez

Dubhe

Phad

Merak

1

Stellar signposts
Arguably the most well-known "starhop" is the one that helps you find the "pole star" Polaris, which sits in Ursa Minor. Use the pointer stars of the Plough asterism, Merak and Dubhe, in Ursa Major.

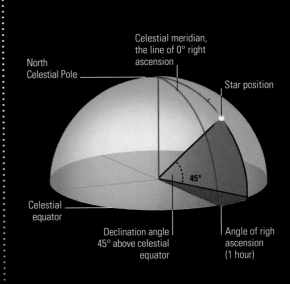

Celestial meridian, the line of 0° right ascension

North Celestial Pole

Star position

Celestial equator

45°

Declination angle 45° above celestial equator

Angle of right ascension (1 hour)

A star's position

THE CHANGING SKY

The night sky changes over time, revealing a panorama of celestial objects. Understanding the sky's movement helps us to predict what will be on show and when.

DAILY SKY MOVEMENTS

The stars seem to move across the sky as the night progresses. This is caused by Earth's rotation on its axis. Once every 24 hours, Earth completes one rotation on its axis relative to the Sun. This is known as a mean solar day. Astronomers also measure the time taken for one rotation of Earth relative to the stars. This is known as a sidereal day, and is slightly shorter than a solar day, at 23 hours 56 minutes and 4 seconds. This means a star will rise four minutes earlier each night. The difference between solar and sidereal days occurs because Earth has moved a little in its orbit around the Sun. The motion of the stars across the sky, over the course of an evening, depends on your location on Earth's surface (see right).

APRIL, 8PM

8 APRIL, 8PM

15 APRIL, 8PM

Moving constellation
Because of a 4-minute difference between the length of a sidereal and a solar day, the constellations move westwards a little from one night to the next.

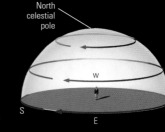

Motion at the north pole
At the north pole, the stars circle in an anticlockwise motion around a po[int] above you. At the s[outh] pole, they move in t[he] opposite direction.

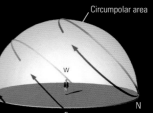

Motion at mid-latitudes
If you are observing from mid-latitudes, will see the stars ris[e] the east and set in t[he] west. Stars that nev[er] set are known as "circumpolar".

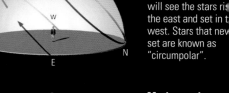

Motion at the equator
If you are standing a[t] the equator, the sta[rs] will rise straight up from the east, pass overhead, and then move straight down to set in the west.

YEARLY SKY MOVEMENTS

Not only do the stars move across the sky during the course of a single night, they also shift slowly around the sky over the course of a year. This means that at one time of the year, a constellation or a region of the sky may be visible when it is dark, say at midnight, while at another time it is hidden behind the Sun. This is because Earth moves around the Sun, so it appears as if the Sun moves against the background night sky.

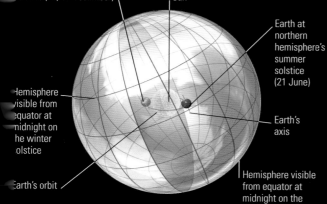

June and December skies
If you were on the equator at midnight in June you would see exactly the opposite half of the celestial sphere

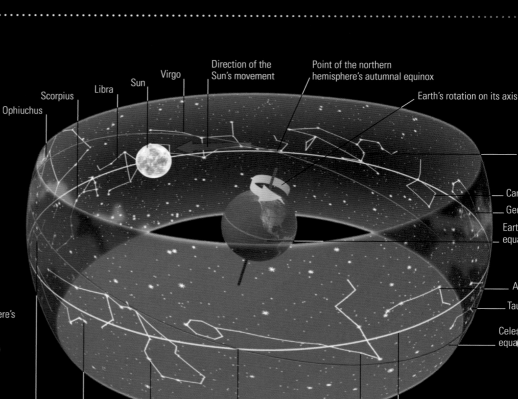

Zodiac
As the Sun appears to move against the background stars over the year it passes through several constellations.

ECLIPSES

As the Moon orbits Earth, it occasionally blocks the Sun's light. The Moon's shadow is cast onto Earth and anyone within the shadow will see a solar eclipse. Standing in the darker part of the Moon's shadow, the umbra, you will see a total solar eclipse with the Sun completely covered by the Moon. If you are in the outer, lighter shadow, known as the penumbra, you will see a partial solar eclipse with only a portion of the Sun obscured by the Moon's disc. If the Moon is too far away it cannot completely cover the Sun and an annular solar eclipse is seen from Earth. Similarly, if the Moon's orbit takes it into the shadow cast into space by Earth, a lunar eclipse occurs. Eclipses require certain precise alignments of the Sun, Earth, and Moon, which is why an eclipse does not occur every time there is a full Moon or a new Moon.

Total lunar eclipse
During a total lunar eclipse, the Moon can often be tinted a wonderful copper-red colour. This is the highlight of one of the greatest spectacles that can be seen in the night sky.

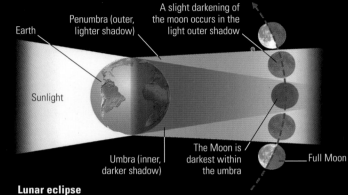

Solar eclipse
Observers in the Moon's umbral shadow see the Sun totally covered by the Moon. Those in the penumbra region only see a partial eclipse – where the Moon partly obscures the Sun's disc.

Lunar eclipse
During a lunar eclipse, the Moon enters the Earth's lighter shadow, the penumbra. It then enters the darker region, or umbra, where it typically goes deep red. Finally, it moves out into the penumbra again.

PLANETARY MOTIONS

Most of the planets can be spotted in the night sky with nothing more than the naked eye. The planets Mercury and Venus are termed the "inferior planets" as they go around the Sun in orbits that are closer to the Sun than Earth's orbit. Because of their proximity to the Sun, they are typically seen low in the sky before sunrise and after sunset. Mars and the planets beyond it are called "superior planets". They can stray far from the Sun in the sky and can be spotted late at night. As most of the planets orbit in roughly the same plane as they go around the Sun they can all be found relatively close to the line of the ecliptic (the path of the Sun on the sky). For more on their locations in the night sky, see the planet locator charts in the Monthly Sky Guides section of this book (pp.20–115).

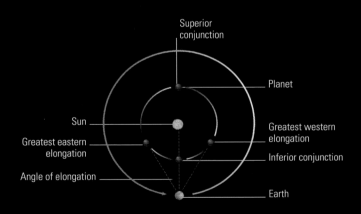

INFERIOR ORBIT

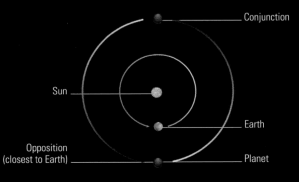

SUPERIOR ORBIT

Mercury and Venus
Many of the planets can be seen very easily with the naked eye. In this photograph, the planets Mercury and Venus sit near each

Planetary positions
These diagrams show several key positions in the orbits of the planets. Inferior planets are best seen near greatest elongation, while superior

GETTING STARTED

Many of the night sky's finest sights can be seen with the naked eye, but telescopes and binoculars allow us to see further and explore celestial objects in more detail.

PLANNING AHEAD

Preparation is the key to an enjoyable night's astronomy under the stars. Start by thinking what you want to look at and the equipment you will need to see it. For example, if you are using equipment that requires batteries make sure they are charged and ready. Additionally, take a good star chart with you (see pp.16–114), so that you know the location of the objects you want to observe. Also think about what to wear when observing. The clearest nights can often be the coldest, so it is crucial to wrap up warm with a windproof jacket as well as waterproof trousers, a warm hat, and stout shoes. If you are travelling out to a remote location, make sure to tell someone where you are going. Remembering these basic things will ensure your observing goes smoothly.

Viewing location
The location from which you observe is as important as the equipment you use. Ideally, head towards a dark sky site, away from sources of light pollution, such as streetlamps and houses.

CHECKLIST

- Warm clothes
- Gloves
- Red light torch
- Notepad and pen
- Any equipment (telescopes, binoculars, etc)
- Compass
- Star chart
- Warm drink
- Blanket or foldable chair

Red light torch
To preserve your night vision, a red light torch is crucial. This can be bought or easily made by covering a regular torch with a red sweet wrapper, secured by an elastic band.

BINOCULARS

Loved by beginners and experts alike, binoculars are a simple, generally inexpensive way to explore the night sky. A good pair of binoculars is capable of showing a huge variety of objects on any given night. With even a small pair the rich star fields of the Milky Way, glittering open star clusters, and the Moon's rugged surface are all wonderful sights. Binoculars come in many different sizes and are defined by two numbers that can often be found marked on their sides. The first number is the number of times the binoculars can magnify a view, and the second number is the size of the main (or objective) lens in millimetres. For example, binoculars that magnify 10x with 50mm diameter lenses are said to be a "ten by fifty" pair of binoculars.

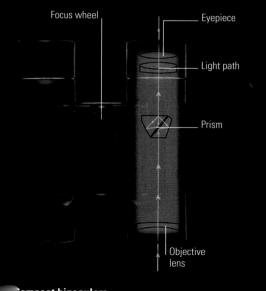

Focus wheel — Eyepiece

Light path

Prism

Objective lens

Eyepiece

Prism

Light path

Objective lens

Compact binoculars
These "roof-prism" binoculars use a design where light passes through a group of close

Standard binoculars
"Porro-prism" binoculars with a folded light path are popular with astronomers as their typically

Large binoculars
Large binoculars can provide stunning views of celestial objects; however, they require

TELESCOPES

Designed to collect light from celestial objects, telescopes also magnify the view and allow us to see objects in more detail. A telescope's crucial specification is its aperture – the size of its main mirror or lens – which is usually measured in millimeters or inches. The larger the main mirror or lens, the more light it will be able to gather. A typical small telescope has an aperture of 10–15cm (4–6in). Telescopes can be held on different types of mounts. These must be sturdy enough to provide a firm, stable platform for the optics and allow the telecope to aim accurately. Equatorial mounts are capable of aligning to the rotation axis of the night sky for simplified tracking. More compact in comparison to equatorial mounts, altitude-azimuth mounts work by moving around 360° (in azimuth) and up and down (in altitude).

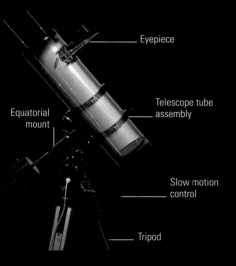

Eyepiece

Telescope tube assembly

Equatorial mount

Slow motion control

Tripod

Newtonian reflector
This simple design, consisting of a tube on a mount and tripod, is ideal for a beginner. The eyepiece is located at the top of the tube and extends from the side.

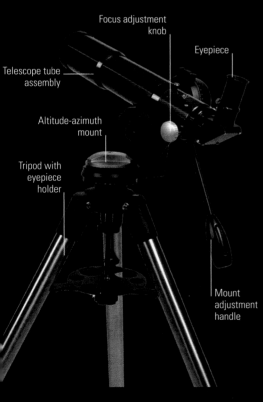

Focus adjustment knob

Eyepiece

Telescope tube assembly

Altitude-azimuth mount

Tripod with eyepiece holder

Mount adjustment handle

Refractor
Refractor telescopes have a classic telescope design, their lenses collect light and produce an image in the eyepiece. They are good for observing a range of celestial objects.

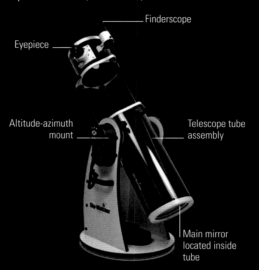

Finderscope

Eyepiece

Altitude-azimuth mount

Telescope tube assembly

Main mirror located inside tube

Dobsonian
The Dobsonian reflector uses a simple altitude-azimuth mount. As they tend to have larger apertures for their price, Dobsonians are excellent deep-sky telescopes.

OBSERVING WITH THE NAKED EYE

There is a wealth of things to see in the night sky using nothing more than the naked eye. For example, meteor showers are best seen by sitting back on a chair and just watching the sky. Similarly, no equipment is needed to marvel at the great expanse of the Milky Way galaxy, or the many stars within it, from a dark sky site. To get the most spectacular naked-eye views of the night sky you must head away from light-polluted towns and cities.

What you can see
Many celestial objects are visible to the naked eye, including the Milky Way, the Andromeda Galaxy, meteors, noctilucent clouds, and the aurorae (shown above).

RECORDING WHAT YOU SEE

There are many ways to record the things you observe when studying the night sky. The simplest is to make a sketch of what you see with the naked eye or through the eyepiece of a telescope or binoculars. Make sure to use a good quality pencil and an artist's sketch pad. To sketch clusters, nebulae, and other deep-sky objects, the best way to begin is to draw the brighter stars first.

Another popular method, though a little more tricky to master, is astrophotography. This involves connecting a camera to your telescope to take pictures. However you record your observations, be sure to note down the observing conditions, the time and date, your name and location, details of the equipment used, and the name of the object observed.

An astro image of the North America Nebula
Astrophotographers create stunning images of the night sky, with the help of sensitive cameras mounted on telescopes. They stack together many individual exposures to create a final detailed image.

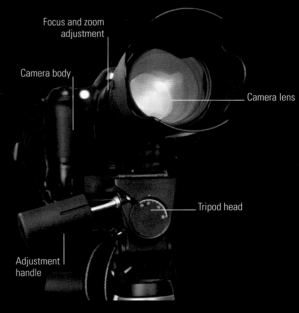

Focus and zoom adjustment

Camera body

Camera lens

Tripod head

Adjustment handle

Digital SLR
These cameras are used by astroimagers as they can be attached to telescopes and their shutters can be left open to gather the faint light from celestial objects.

MONTHLY
SKY GUIDES

• •

During the course of a year our view of the
night sky changes from month to month as
Earth orbits the Sun. Some constellations
are always in the sky, while others appear
and disappear over different regions. The
following monthly sky guides help you to
recognize patterns and track changes in
the northern and southern hemispheres.

The aurorae
The Northern and Southern lights, or the aurorae, occur in high
latitudes of both hemispheres such as Wapusk National Park in
Canada, as seen here. They cause spectacular displays that often
last for hours and end with a ribbon-like rippling effect.

USING THE SKY GUIDES

This month-by-month guide features charts that show the whole night sky as it appears from different locations. These pages explain how to get the most from the information the charts give.

MONTHLY OVERVIEWS

For each month of the year, a double-page spread outlines the different celestial phenomena in the sky. These include bright stars, constellations, deep-sky objects, and meteor showers. The constellation box on these pages discusses a key constellation in detail, pointing out its bright stars.

These pages also feature planet-locator charts, which show the band of sky that lies on either side of the ecliptic, where the planets appear. These charts should be used in conjunction with the information supplied in the following highlights pages, whole-sky charts, and the Almanac.

Uranus and Neptune
The magnified insets of the main chart show Uranus and Neptune, the two outermost planets, as they move relatively slowly through our sky.

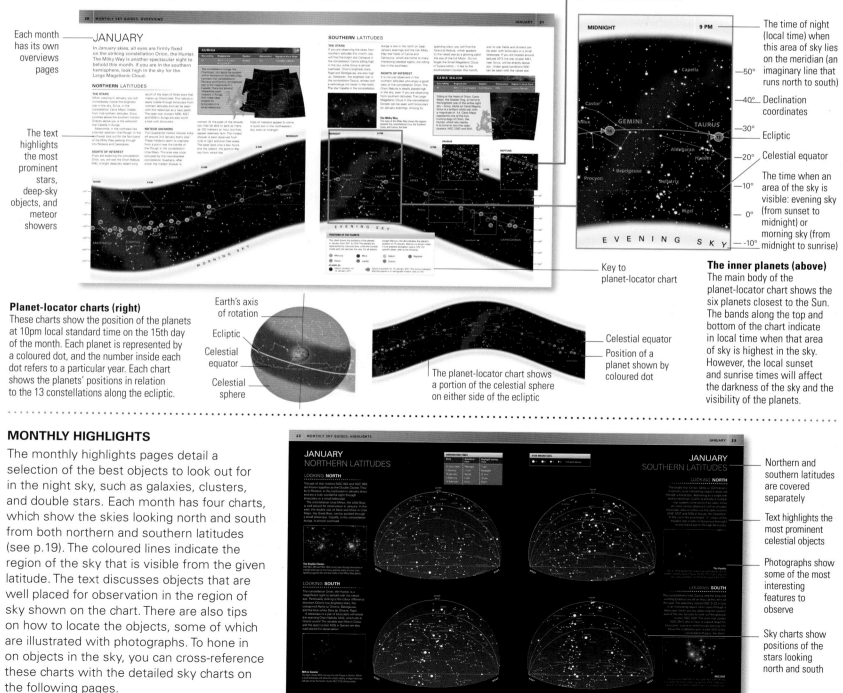

Each month has its own overviews pages

The text highlights the most prominent stars, deep-sky objects, and meteor showers

The time of night (local time) when this area of sky lies on the meridian (an imaginary line that runs north to south)

Declination coordinates

Ecliptic

Celestial equator

The time when an area of the sky is visible: evening sky (from sunset to midnight) or morning sky (from midnight to sunrise)

Key to planet-locator chart

The inner planets (above)
The main body of the planet-locator chart shows the six planets closest to the Sun. The bands along the top and bottom of the chart indicate in local time when that area of sky is highest in the sky. However, the local sunset and sunrise times will affect the darkness of the sky and the visibility of the planets.

Planet-locator charts (right)
These charts show the position of the planets at 10pm local standard time on the 15th day of the month. Each planet is represented by a coloured dot, and the number inside each dot refers to a particular year. Each chart shows the planets' positions in relation to the 13 constellations along the ecliptic.

Earth's axis of rotation

Ecliptic

Celestial equator

Celestial sphere

Celestial equator

Position of a planet shown by coloured dot

The planet-locator chart shows a portion of the celestial sphere on either side of the ecliptic

MONTHLY HIGHLIGHTS

The monthly highlights pages detail a selection of the best objects to look out for in the night sky, such as galaxies, clusters, and double stars. Each month has four charts, which show the skies looking north and south from both northern and southern latitudes (see p.19). The coloured lines indicate the region of the sky that is visible from the given latitude. The text discusses objects that are well placed for observation in the region of sky shown on the chart. There are also tips on how to locate the objects, some of which are illustrated with photographs. To hone in on objects in the sky, you can cross-reference these charts with the detailed sky charts on the following pages.

Northern and southern latitudes are covered separately

Text highlights the most prominent celestial objects

Photographs show some of the most interesting features to observe

Sky charts show positions of the stars looking north and south

THE WHOLE-SKY CHARTS

As well as the monthly highlights and overviews pages, there are two whole-sky charts for every month. These charts show the position of the stars at 10pm local time on the 15th day of the month, for both the northern and southern hemispheres. They project the half of the celestial sphere that would be visible to you without any obstruction on the horizon. To use the whole-sky charts, first use the world map (bottom right) to find the coloured latitude line, which is closest to your observing location. Then turn to the chart for the month you are observing in. Next, look for the horizon line that is coloured the same as the latitude line that is closest to your location. The sky plotted within the horizon's boundaries is visible from your location during that month for the times shown. Now turn to the appropriate month and position yourself and the chart (see right).

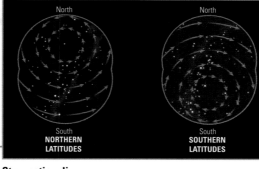

Star-motion diagrams
These diagrams show the direction in which the stars appear to move in the sky as the night progresses. Stars near the celestial equator appear to move from east to west, while circumpolar stars circle around the celestial poles without setting.

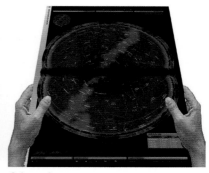

Orientation
To view the northern sky, turn northwards and hold the map flat with the label NORTH closest to you. The coloured northern horizon line on the chart corresponds to the horizon in front of you. To view the south, turn yourself and the map around.

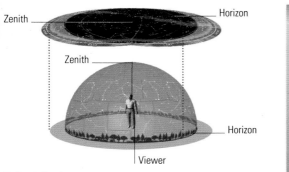

Zenith — Horizon

Zenith

Horizon

Viewer

Celestial sphere
Each whole-sky chart shows an area greater than half a celestial sphere because it combines three different projections of the night sky, as seen from three different latitudes. Each month the sky charts show the sky as it appears from 60°–20°N on the northern latitudes chart and from 0°–40°S on the southern latitudes chart.

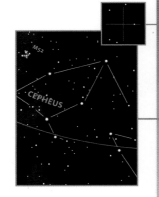

Horizons and zeniths
The stars shown near the centre of each chart are close to the point directly overhead, known as the zenith, while stars near the chart's edge appear close to the horizon. Colour-coded lines and crosses are used to identify the horizon and zenith for each of the three latitude projections on each monthly whole-sky chart.

JANUARY | NORTHERN LATITUDES

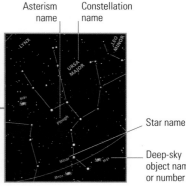

Asterism name   Constellation name

Star name

Deep-sky object name or number

Main features
Besides showing the band of the Milky Way galaxy, the whole-sky charts also highlight many stars, constellations, deep-sky objects, asterisms, and the path of the Sun – also known as the ecliptic.

OBSERVATION TIMES

Date	Standard time	Daylight-saving time
15 December	Midnight	1 am
1 January	11 pm	Midnight
15 January	10 pm	11 pm
1 February	9 pm	10 pm
15 February	8 pm	9 pm

Observation times
Each chart shows the sky as it appears at 10pm local standard time mid month. However, this view can also be seen at other times of the month, as well as one hour later when local daylight saving time is in use. These times range from midnight in the middle of the previous month to 8pm in the middle of the next month.

STAR MAGNITUDES

● -1 ● 0 ● 1 • 2 · 3 · 4 · 5 ⊙ Variable star

Star magnitude
This panel shows the approximate magnitudes of the stars plotted on the whole-sky chart. In astronomy, the brighter a star the lower its "magnitude" value; bright objects may even have a negative value.

DEEP-SKY OBJECTS

🌀 Galaxy   ❋ Globular cluster   ✦ Open cluster   ☁ Diffuse nebula   ⬭ Planetary nebula

Deep-sky objects
This panel displays the symbols for deep-sky objects, including galaxies, clusters, and nebulae shown on the whole-sky charts.

POINTS OF REFERENCE

Horizons 60°N 40°N 20°N   Zeniths + 60°N + 40°N + 20°N Ecliptic

Points of reference
To help you remember which horizon line or zenith marker applies to your location, this bar shows the different colours that correspond to the different latitudes.

60°N
40°N
20°N
0°
20°S
40°S

Lines of latitude
Use this map to find the coloured latitude line that is closest to your location. This is the colour of the line you need to find on the sky charts (see above). Note that a 10° difference in latitude has little effect on the stars that can be seen.

JANUARY

In January skies, all eyes are firmly fixed on the striking constellation Orion, the Hunter. The Milky Way is another spectacular sight to behold this month. If you are in the southern hemisphere, look high in the sky for the Large Magellanic Cloud.

NORTHERN LATITUDES

THE STARS

When viewing in January, you will immediately notice the brightest star in the sky, Sirius, in the constellation Canis Major. Visible from mid-northern latitudes, Sirius twinkles above the southern horizon. Directly above you is the yellowish star Capella in Auriga.

Meanwhile, in the northeast lies a familiar asterism, the Plough. In the northwest look out for the faint band of the Milky Way passing through into Perseus and Cassiopeia.

SIGHTS OF INTEREST

If you are exploring the constellation Orion, you will see the Orion Nebula, M42, a bright deep-sky object lying south of the chain of three stars that makes up Orion's belt. This nebula is easily visible through binoculars from northern latitudes and can be seen with the naked eye as a hazy patch. The open star clusters M36, M37, and M38 in Auriga are also worth a look with binoculars.

METEOR SHOWERS

The Quadrantid meteor shower kicks off around 3–4 January every year. These meteors seem to originate from a point near the handle of the Plough in the constellation Ursa Major. This area was once occupied by the now-obsolete constellation Quadrans, after which the meteor shower is named. At the peak of the shower, you may be able to spot as many as 100 meteors an hour, but they appear relatively faint. The meteor shower is best observed from rural or light pollution-free areas. Their peak lasts only a few hours and their radiant, the point in the sky from which the trails of meteors appear to come, is quite low in the northeastern sky, even at midnight.

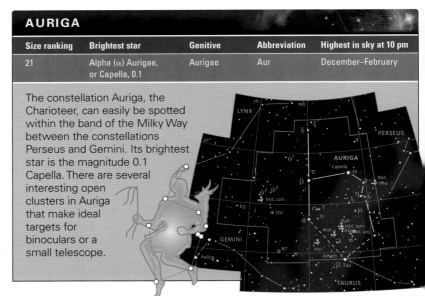

AURIGA				
Size ranking	Brightest star	Genitive	Abbreviation	Highest in sky at 10 pm
21	Alpha (α) Aurigae, or Capella, 0.1	Aurigae	Aur	December–February

The constellation Auriga, the Charioteer, can easily be spotted within the band of the Milky Way between the constellations Perseus and Gemini. Its brightest star is the magnitude 0.1 Capella. There are several interesting open clusters in Auriga that make ideal targets for binoculars or a small telescope.

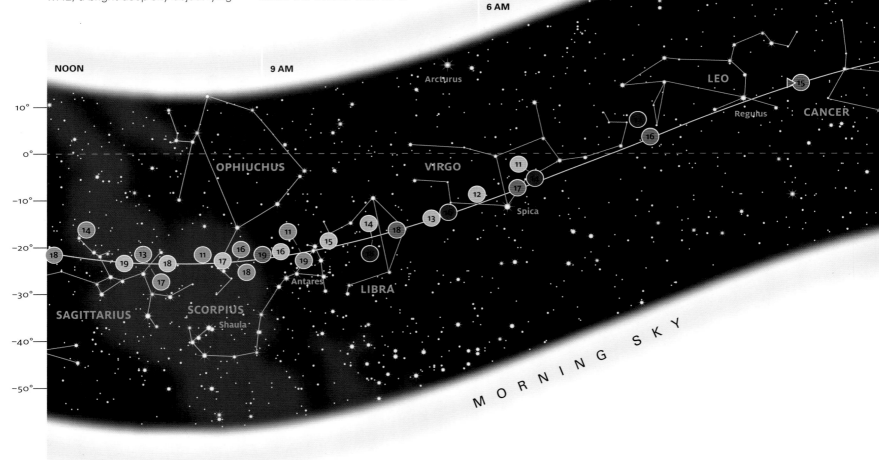

SOUTHERN LATITUDES

THE STARS

If you are observing the skies from southern latitudes this month, you will find the bright star Canopus in the constellation Carina sitting high in the sky, while Sirius is almost overhead. Orion's brightest stars, Rigel and Betelgeuse, are also high up. Aldebaran, the brightest star in the constellation Taurus, shines with a red-orange tint lower in the north. The star Capella in the constellation Auriga is low in the north on clear January evenings and the rich Milky Way star fields of Carina and Centaurus, which are home to many interesting celestial sights, are sitting low in the southeast.

SIGHTS OF INTEREST

It is not just observers in the northern latitudes who enjoy a good view of the constellation Orion. The Orion Nebula is ideally placed high in the sky, even if you are observing from southern latitudes. The Large Magellanic Cloud in the constellation Dorado can be seen with binoculars on January evenings. Among its sparkling stars, you will find the Tarantula Nebula, which appears to the naked eye as a glowing patch the size of the full Moon. Do not forget the Small Magellanic Cloud in Tucana either – it lies to the southwestern horizon this month, and its star fields and clusters can be seen with binoculars or a small telescope. If you are located around latitude 20°S the star cluster M41, near Sirius, will be directly above you. Under good conditions M41 can be seen with the naked eye.

The Milky Way

This view of the Milky Way shows the regions towards the constellations Crux, the Southern Cross, and Carina, the Keel.

MIDNIGHT

9 PM

CANIS MAJOR

Size ranking	Brightest star	Genitive	Abbreviation	Highest in sky at 10 pm
43	Alpha (α) Canis Majoris, or Sirius, -1.4	Canis Majoris	CMa	January–February

Sitting at the heels of Orion, Canis Major, the Greater Dog, is home to the brightest star of the entire night sky – Sirius, Alpha (α) Canis Majoris. Sirius is a brilliant white star with a magnitude of -1.4. Canis Major represents one of the two hunting dogs of Orion, the Hunter, which sits nearby. It is home to two fine open clusters, NGC 2362 and M41.

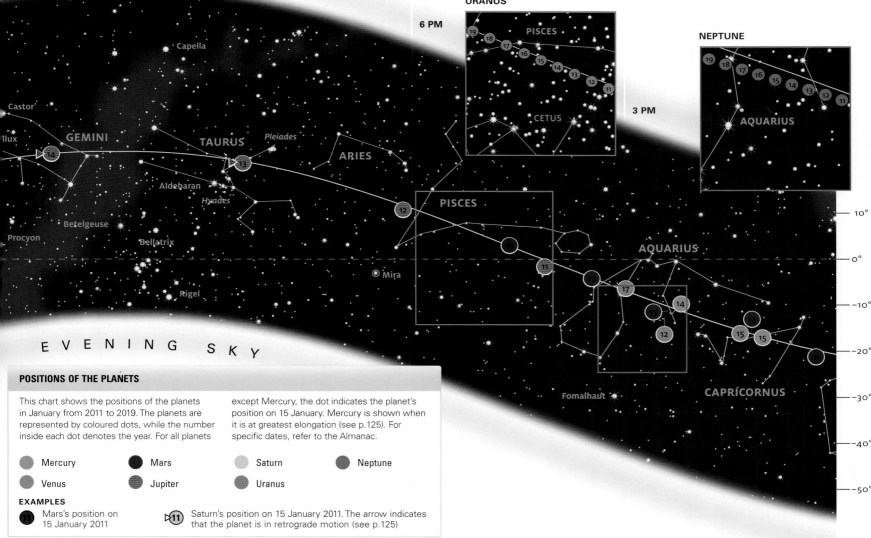

URANUS

6 PM

3 PM

NEPTUNE

EVENING SKY

POSITIONS OF THE PLANETS

This chart shows the positions of the planets in January from 2011 to 2019. The planets are represented by coloured dots, while the number inside each dot denotes the year. For all planets except Mercury, the dot indicates the planet's position on 15 January. Mercury is shown when it is at greatest elongation (see p.125). For specific dates, refer to the Almanac.

- Mercury
- Venus
- Mars
- Jupiter
- Saturn
- Uranus
- Neptune

EXAMPLES

Mars's position on 15 January 2011

Saturn's position on 15 January 2011. The arrow indicates that the planet is in retrograde motion (see p.125)

JANUARY
NORTHERN LATITUDES

OBSERVATION TIMES		
Date	Standard time	Daylight-saving time
15 December	Midnight	1 am
1 January	11 pm	Midnight
15 January	10 pm	11 pm
1 February	9 pm	10 pm
15 February	8 pm	9 pm

LOOKING **NORTH**

The pair of star clusters NGC 884 and NGC 869 are known together as the Double Cluster. They lie in Perseus, in the northwest in January skies and are a truly wonderful sight through binoculars or a small telescope.

The constellation Ursa Minor, the Little Bear, is well placed for observation in January. In the east, the double star of Alcor and Mizar in Ursa Major, the Great Bear, can be spotted through a small telescope. Capella, in the constellation Auriga, is almost overhead.

The Double Cluster
Both NGC 884 and NGC 869 can be seen through binoculars or a small telescope as two fuzzy patches made of many stars, sparkling against the rich star fields of the Milky Way galaxy.

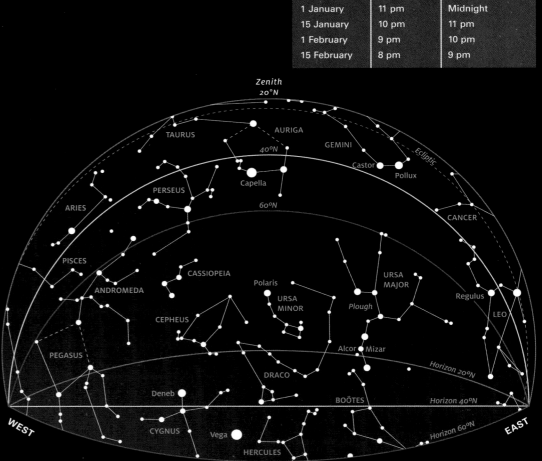

LOOKING **SOUTH**

The constellation Orion, the Hunter, is a magnificent sight to behold with the naked eye. Particularly striking is the colour difference between Orion's two brightest stars, the orange-red Alpha (α) Orionis, Betelgeuse, and the blue-white Beta (β) Orionis, Rigel.

A telescope or a pair of binoculars will reveal the stunning Orion Nebula, M42, which sits in Orion's sword. The variable star Mira in Cetus and the open cluster M35 in Gemini are also well placed for observation.

M35 in Gemini
The open cluster M35 sits near the star Propus in Gemini. While a small telescope will show the cluster clearly, a larger telescope will also show the fainter cluster NGC 2158 sitting nearby.

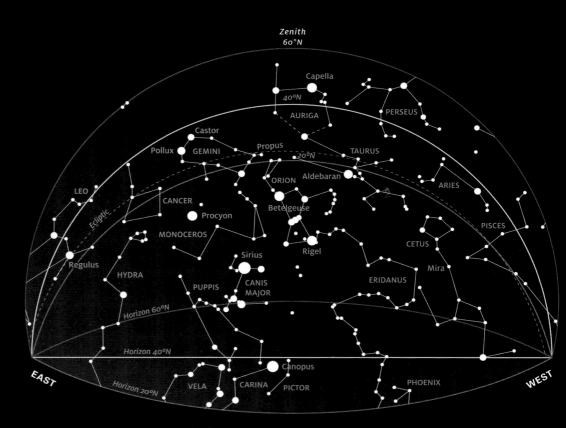

STAR MAGNITUDES

● -1　● 0　● 1　● 2　• 3 and above

Zenith
40°S

ERIDANUS

20°S

CANIS
MAJOR

Rigel

Sirius

PUPPIS

ORION 20°

MONOCEROS

Betelgeuse

CETUS

TAURUS

Aldebaran

Procyon

HYDRA

CANCER

GEMINI

Pollux

Ecliptic

ARIES

AURIGA

Castor

Capella

Horizon 40°S

Regulus

PISCES

PERSEUS

LEO

Horizon 20°S

URSA
MAJOR

WEST

ANDROMEDA

CASSIOPEIA

Plough

EAST

Horizon 0°

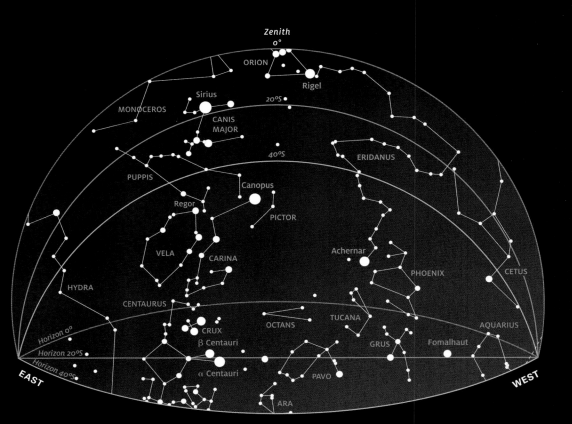

Zenith
0°

ORION

Rigel

Sirius

20°S

MONOCEROS

CANIS
MAJOR

40°S

ERIDANUS

PUPPIS

Canopus

Regor

PICTOR

Achernar

PHOENIX

CETUS

VELA

CARINA

HYDRA

CENTAURUS

OCTANS

TUCANA

AQUARIUS

Horizon 0°

CRUX

β Centauri

GRUS

Fomalhaut

Horizon 20°S

Horizon 40°S

α Centauri

PAVO

EAST

ARA

WEST

LOOKING **NORTH**

The bright star Castor, Alpha (α) Geminorum, in Gemini, is an interesting target if observed through a telescope. Appearing as a single star to the naked eye, Castor is actually a multiple star system composed of six stars, three of which can be observed with an amateur telescope. Also on show are the open clusters M36, M37, and M38 in Auriga, the Charioteer. In the west, the prominent "V" shape of the Hyades star cluster in Taurus is a fine sight to the naked eye or through binoculars.

The Hyades
The "V" of the Hyades star cluster is an unmistakable sight; its members form the head of Taurus, the Bull. The bright orange-red star nearby, Aldebaran, is not actually a part of the cluster.

LOOKING **SOUTH**

The constellations Vela, Carina, and the long and winding Eridanus are all on show at this time of the year. The planetary nebula NGC 3132 in Vela is an interesting object when seen through a telescope. While you are observing the eastern part of the sky, be sure to seek out the globular cluster NGC 3201. The open star cluster NGC 2547, also in Vela, is a good target for binoculars. Use a small telescope to bring into focus the scattered open cluster M47 in the constellation Puppis, the Stern.

NGC 2547
The gorgeous magnitude 4.7 star cluster NGC 2547 sits very close to the star Regor, Gamma (γ) Velorum, in the constellation Vela. It is a fine sight through binoculars or a small telescope.

JANUARY | NORTHERN LATITUDES

LOOKING NORTH

OBSERVATION TIMES		
Date	**Standard time**	**Daylight-saving time**
15 December	Midnight	1 am
1 January	11 pm	Midnight
15 January	10 pm	11 pm
1 February	9 pm	10 pm
15 February	8 pm	9 pm

JANUARY | NORTHERN LATITUDES

LOOKING SOUTH

WEST

SOUTHWEST

SOUTH

SOUTHEAST

EAST

WEST

STAR MOTION

North

South

STAR MAGNITUDES

* -1 * 0 * 1 • 2 • 3 . 4 . 5 ⊛ Variable star

DEEP-SKY OBJECTS

🌀 Galaxy   ⊙ Globular cluster   ✳ Open cluster   ✺ Diffuse nebula   ⊙ Planetary nebula

POINTS OF REFERENCE

Horizons | 60°N | 40°N | 20°N   Zeniths | 60°N | 40°N | 20°N | Ecliptic

Constellations and objects labeled

SCULPTOR

PISCES

ARIES

PERSEUS

M45 (Pleiades)

AURIGA

M38

M36

M37

M35

GEMINI

Castor

Pollux

CANCER

M44

M67

LEO

Regulus

SEXTANS

HYDRA

Ecliptic

CETUS

Mira

TAURUS

Aldebaran

Hyades

ORION

M1

Bellatrix

Betelgeuse

Rigel

M42

MONOCEROS

CANIS MINOR

Procyon

M50

M47

M46

M48

M41

Sirius

CANIS MAJOR

Adhara

M93

PYXIS

ANTLIA

VELA

PUPPIS

Canopus

CARINA

PICTOR

DORADO

LMC

RETICULUM

HOROLOGIUM

CAELUM

COLUMBA

LEPUS

ERIDANUS

FORNAX

PHOENIX

JANUARY | SOUTHERN LATITUDES

STAR MAGNITUDES

✦	-1
✦	0
✦	1
•	2
•	3
•	4
•	5
⊛	Variable star

DEEP-SKY OBJECTS

🌀	Galaxy
⬤	Globular cluster
✳	Open cluster
✺	Diffuse nebula
◉	Planetary nebula

POINTS OF REFERENCE

Horizons	0°	20°S	40°S
Zeniths	✛ 0°	✛ 20°S	✛ 40°S
			Ecliptic

OBSERVATION TIMES

Date	Standard time	Daylight-saving time
15 December	Midnight	1 am
1 January	11 pm	Midnight
15 January	10 pm	11 pm
1 February	9 pm	10 pm
15 February	8 pm	9 pm

LOOKING NORTH

JANUARY | SOUTHERN LATITUDES

LOOKING SOUTH

WEST

SOUTHWEST

SOUTH

SOUTHEAST

EAST

WEST

AQUARIUS

CETUS

FORNAX

ERIDANUS

LEPUS

COLUMBA

CAELUM

CANIS MAJOR

Adhara

M41

PUPPIS

M93

PYXIS

ANTLIA

NGC 3132

HYDRA

CRATER

CORVUS

VELA

NGC 3201

CARINA

NGC 2547

Canopus

PICTOR

DORADO

LMC

RETICULUM

HOROLOGIUM

Achernar

SCULPTOR

PISCIS AUSTRINUS

Fomalhaut

PHOENIX

HYDRUS

NGC 104

SMC

MENSA

VOLANS

CHAMAELEON

MUSCA

Acrux

Becrux

Gacrux

CRUX

β Centauri

α Centauri

CENTAURUS

NGC 5139

LUPUS

CIRCINUS

TRIANGULUM AUSTRALE

APUS

OCTANS

PAVO

ARA

INDUS

TUCANA

GRUS

MICROSCOPIUM

STAR MOTION

North

South

STAR MAGNITUDES

· -1 ★ 0 · 1 · 2 · 3 · 4 · 5 ⊛ Variable star

DEEP-SKY OBJECTS

🌀 Galaxy ✦ Globular cluster ✳ Open cluster ☁ Diffuse nebula ◎ Planetary nebula

POINTS OF REFERENCE

Horizons | 0° 20°S 40°S

Zeniths | + 0° + 20°S + 40°S

| Ecliptic

FEBRUARY

To get your bearings this month, look out for the bright stars Castor and Pollux in Gemini from northern latitudes. If you are observing from the southern hemisphere, the constellations Carina, Puppis, and Vela can be seen high in the sky.

GEMINI

Size ranking	Brightest stars	Genitive	Abbreviation	Highest in sky at 10 pm
30	Beta (β) Geminorum, 1.15 Alpha (α) Geminorum, 1.6	Geminorum	Gem	January–February

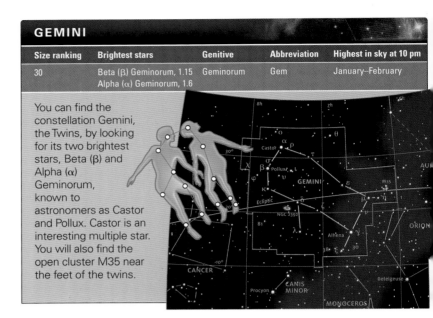

You can find the constellation Gemini, the Twins, by looking for its two brightest stars, Beta (β) and Alpha (α) Geminorum, known to astronomers as Castor and Pollux. Castor is an interesting multiple star. You will also find the open cluster M35 near the feet of the twins.

NORTHERN LATITUDES

THE STARS

If you are observing the sky from mid-northern latitudes, you will find the constellation Gemini almost overhead. South of Gemini lies the sparkling Winter Triangle formed by the bright stars Sirius in Canis Major, Betelgeuse in Orion, and Procyon in Canis Minor. The four constellations Taurus, Auriga, Perseus, and the W-shaped Cassiopeia are also on show this month. In the eastern sky the constellation Leo, the Lion, is visible, while the familiar figure of the Plough asterism sits nearby in the northeast.

SIGHTS OF INTEREST

Close to the feet of Gemini, the Twins, you will find M35, a large open star cluster that can be easily spotted with a pair of binoculars.

The Winter Triangle
In the northern winter night sky, look out for the stars of the Winter Triangle asterism – Sirius (centre bottom), the orange-red Betelgeuse (top), and Procyon (top left).

The wonderful Beehive Cluster, also known as M44 or Praesepe is a great sight through a small telescope. It lies in the nearby constellation Cancer and appears wider than the full Moon. Meanwhile, you will find the Milky Way running through the constellation Monoceros, home to many open star clusters. If you are observing with a pair of binoculars, look out for the star cluster NGC 2244. It is also an interesting target for a small telescope, and sits between the stars Betelgeuse and Procyon.

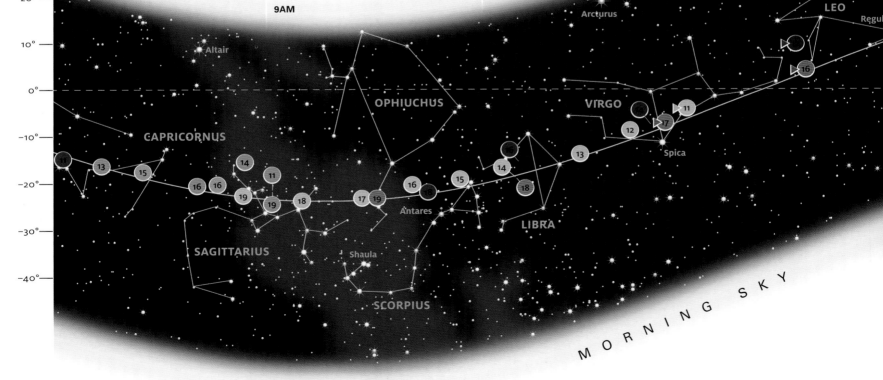

SOUTHERN LATITUDES

THE STARS

If you are observing from southern latitudes you will find two stellar beacons shining away high in the sky. These are the two brightest stars in the night sky – Sirius and Canopus. Two stunning constellations Crux, the Southern Cross, and Centaurus, the Centaur, are also visible. A little way above them, you will find the deceptive False Cross, sometimes mistaken for the true Southern Cross. The False Cross is formed by four stars in Vela and Carina.

At this time of the year, the two main stars of Gemini, Castor and Pollux, can be found sitting in the north. The constellations Orion and Taurus are also on show. In the south you will still find the Large and Small Magellanic Clouds. Meanwhile Leo, the Lion, is steadily rising in the northeast.

SIGHTS OF INTEREST

The regions in and around the Milky Way should be explored as they contain many star clusters, such as M46, M47, NGC 2451, and NGC 2477, which can be found in Puppis. Also seek out the star clusters IC 2391 and IC 2395 in Vela and NGC 2516 in Carina.

PUPPIS

Size ranking	Brightest star	Genitive	Abbreviation	Highest in sky at 10 pm
20	Zeta (ζ) Puppis, or Naos, 2.2	Puppis	Pup	January–February

The constellation Puppis, the Stern, is home to the open star clusters M46 and M47, which can be seen with a pair of binoculars. Puppis sits in the night sky just north of the bright star Canopus and is wedged between the constellations Vela, Carina, and Canis Major.

The South Celestial Pole
Find the South Celestial Pole by intersecting two imaginary lines: one, an extension of the long axis of Crux, and the other at right angles to the line joining Alpha and Beta Centauri.

URANUS

9PM

MIDNIGHT

6PM

3PM

EVENING SKY

NEPTUNE

POSITIONS OF THE PLANETS

This chart shows the positions of the planets in February from 2011 to 2019. The planets are represented by coloured dots, while the number inside each dot denotes the year. For all planets except Mercury, the dot indicates the planet's position on 15 February. Mercury is shown when it is at greatest elongation (see p.125). For specific dates, refer to the Almanac.

- ● Mercury
- ● Mars
- ● Saturn
- ● Neptune
- ● Venus
- ● Jupiter
- ● Uranus

EXAMPLES

● Mars's position on 15 February 2011

🔁11 Saturn's position on 15 February 2011. The arrow indicates that the planet is in retrograde motion (see p.125)

FEBRUARY
NORTHERN LATITUDES

OBSERVATION TIMES		
Date	**Standard time**	**Daylight-saving time**
15 January	Midnight	1 am
1 February	11 pm	Midnight
15 February	10 pm	11 pm
1 March	9 pm	10 pm
15 March	8 pm	9 pm

LOOKING **NORTH**

In February, the three prominent open clusters in the northwest M36, M37, and M38 in Auriga are a must-see. Through a telescope, each cluster appears like grains of sugar scattered against the black sky; a pair of binoculars show the clusters as grey smudges.

Other objects to be spotted with binoculars are the galaxy M81 in Ursa Major, the Great Bear, and the line of stars known as Kemble's Cascade in Camelopardalis, the Giraffe, which lies close to Cassiopeia and Perseus.

Kemble's Cascade
Sitting close to the halfway point between the bright star Capella, in Auriga, and Gamma (γ)Cassiopeiae, Kemble's Cascade is best observed with a pair of binoculars.

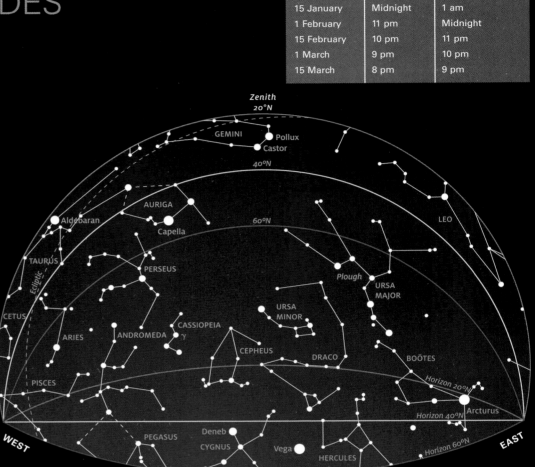

LOOKING **SOUTH**

The open cluster M41 lies just south of the bright star Sirius in Canis Major, the Greater Dog. It is worth a look if you are enjoying the more obvious sights of Orion nearby. M41 is clearly visible through binoculars or a small telescope.

To the east, the cluster NGC 2244 in Monoceros, the Unicorn, is good through binoculars and small telescopes. Also worth finding is M1 in Taurus in the west. A large telescope reveals its elliptical shape, while the biggest telescopes reveal even more details.

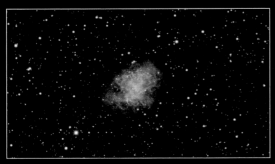

M1 in Taurus
Also called the Crab Nebula, M1 is a supernova remnant that was formed by the explosion of a massive star. About 6,500 light-years from Earth, it appears as a faintly glowing patch in the sky.

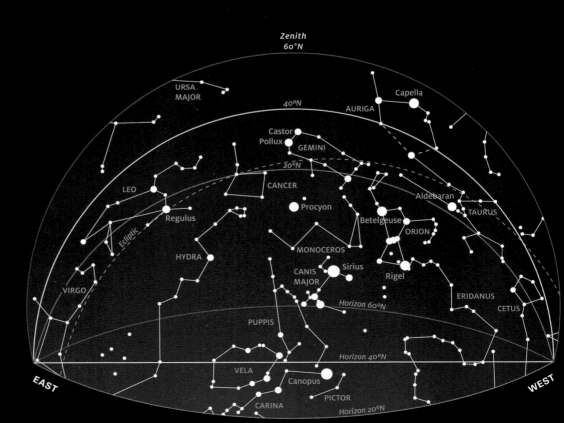

FEBRUARY
SOUTHERN LATITUDES

STAR MAGNITUDES
● -1 ● 0 ● 1 ● 2 • 3 and above

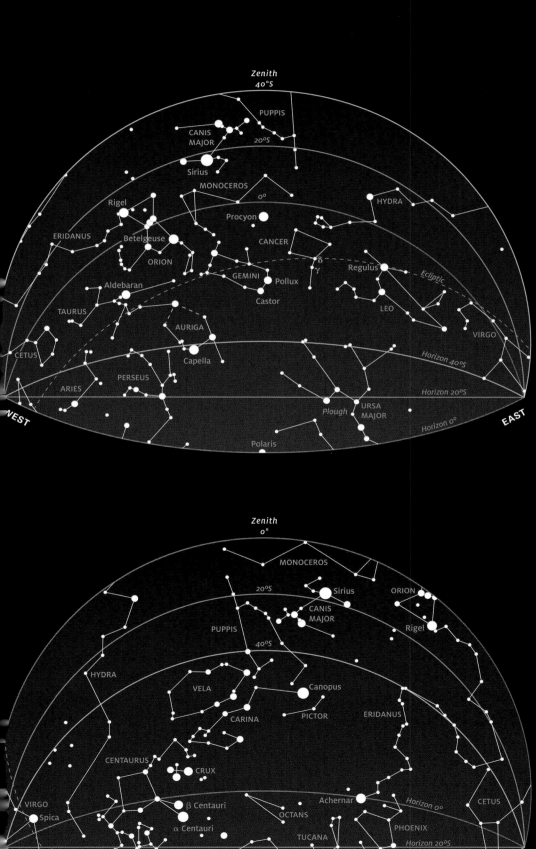

LOOKING **NORTH**

In the constellation Cancer, the Crab, look out for the wonderful star cluster M44, also known as the Beehive Cluster or Praesepe. Easy to locate, it sits at the very centre of the constellation, close to the stars Gamma (γ) and Delta (δ) Cancri. M44 appears as a misty patch to the naked eye from a dark sky location. Also on show, and best seen with a telescope, are the interesting spiral galaxies M65 and M66 in Leo, the Lion. These galaxies appear as elliptical smudges through a telescope.

M44 in Cancer

Appearing as a glittering collection of stars, the magnitude 3.7 open cluster M44 is a beautiful sight through binoculars. It is an ideal target for amateur deep-sky astrophotographers.

LOOKING **SOUTH**

This is a good time to observe and admire the rich star fields of the Milky Way stretching across the southern skies. To scan these stars you only need your eyes or a pair of binoculars. Look out for the Milky Way running through the constellations Crux, Centaurus, Musca, and Carina in the east. Be sure to observe the Coalsack Nebula, a distinctive dark patch close to the stars of Crux, the Southern Cross. This dark nebula is a cloud of dust and gas about 600 light-years away.

IC 2602

This magnitude 1.9 open cluster lying in the constellation Carina is known to astronomers as the Southern Pleiades. It is visible

FEBRUARY | NORTHERN LATITUDES

STAR MAGNITUDES

DEEP-SKY OBJECTS

POINTS OF REFERENCE

LOOKING NORTH

OBSERVATION TIMES		
Date	Standard time	Daylight-saving time
15 January	Midnight	1 am
1 February	11 pm	Midnight
15 February	10 pm	11 pm
1 March	9 pm	10 pm
15 March	8 pm	9 pm

FEBRUARY | NORTHERN LATITUDES

LOOKING SOUTH

WEST

EAST

SOUTHEAST

SOUTH

SOUTHWEST

WEST

STAR MOTION

North

South

STAR MAGNITUDES

-1 · 0 · 1 · 2 · 3 · 4 · 5 · Variable star

DEEP-SKY OBJECTS

Galaxy · Globular cluster · Open cluster · Diffuse nebula · Planetary nebula

POINTS OF REFERENCE

Horizons · 60°N · 40°N · 20°N · Zeniths · 60°N · 40°N · 20°N · Ecliptic

Constellations and objects labeled: CETUS, Mira, ERIDANUS, FORNAX, TAURUS, Hyades, Aldebaran, ORION, Bellatrix, Betelgeuse, Rigel, M42, M43, LEPUS, COLUMBA, CAELUM, PICTOR, DORADO, AURIGA, M38, M36, M37, M35, M1, GEMINI, Castor, Pollux, CANIS MINOR, Procyon, MONOCEROS, NGC 2244, Sirius, CANIS MAJOR, Adhara, M50, M41, Canopus, PUPPIS, M46, M47, M93, CARINA, VOLANS, CANCER, M44, M67, M48, HYDRA, PYXIS, VELA, ANTLIA, LEO, Regulus, Ecliptic, SEXTANS, CRATER, CORVUS, M104, VIRGO, M87

FEBRUARY | SOUTHERN LATITUDES

LOOKING NORTH

STAR MAGNITUDES

✦	✦	✶	✴	·	·	·	⊛
–1	0	1	2	3	4	5	Variable star

DEEP-SKY OBJECTS

🌀 Galaxy
⬤ Globular cluster
✳ Open cluster
✺ Diffuse nebula
◉ Planetary nebula

POINTS OF REFERENCE

| **Horizons** | 0° | 20°S | 40°S |
| **Zeniths** | + 0° | + 20°S | + 40°S | Ecliptic |

OBSERVATION TIMES

Date	Standard time	Daylight-saving time
15 January	Midnight	1 am
1 February	11 pm	Midnight
15 February	10 pm	11 pm
1 March	9 pm	10 pm
15 March	8 pm	9 pm

FEBRUARY | SOUTHERN LATITUDES

WEST

CETUS

SCULPTOR

FORNAX

ERIDANUS

PHOENIX

SOUTHWEST

HOROLOGIUM

Achernar

GRUS

STAR MOTION

North

South

CAELUM

COLUMBA

LEPUS

CANIS MAJOR

Canopus

DORADO

RETICULUM

HYDRUS

MENSA

SMC

NGC 104

TUCANA

INDUS

PICTOR

LMC

M41

Adhara

PUPPIS

PYXIS

CARINA

VOLANS

OCTANS

LOOKING SOUTH

SOUTH

APUS

PAVO

IC 2602

CHAMAELEON

M93

VELA

CRUX

Acrux

MUSCA

Gacrux

Bccrux Coalsack
Nebula

β Centauri

TRIANGULUM
AUSTRALE

ARA

ANTLIA

α Centauri

CIRCINUS

HYDRA

NGC 5139

NORMA

CENTAURUS

CRATER

LUPUS

SOUTHEAST

CORVUS

M104

M83

VIRGO

Spica

EAST

WEST

POINTS OF REFERENCE

Planetary

Diffuse

Globular

Open

DEEP-SKY OBJECTS

Variable

STAR MAGNITUDES

MARCH

As the nights grow shorter in the northern hemisphere, the bright winter constellations move towards the west. In the southern hemisphere, however, the nights are getting longer, bringing many fine celestial objects into view.

NORTHERN LATITUDES

THE STARS

Look north on March evenings and you will see the "Sickle" asterism, which makes up the head of Leo, the Lion. To its right is the less conspicuous constellation Cancer. Below this region are the rather faint and sparse constellations Sextans, Crater, and Hydra. The most notable star in this part of the sky is Alphard, lying in the constellation Hydra. Appropriately Alphard means "the solitary one".

Sitting high in the northeast is the reassuringly familiar shape of the Plough asterism, with its handle arching down towards the bright star Arcturus in Boötes. A little way away and closer to the horizon lies the star Spica in the constellation Virgo. Also

The Sickle of Leo
The asterism known as the "Sickle", made from the stars of the head of Leo, is a useful celestial signpost for navigating March's night skies.

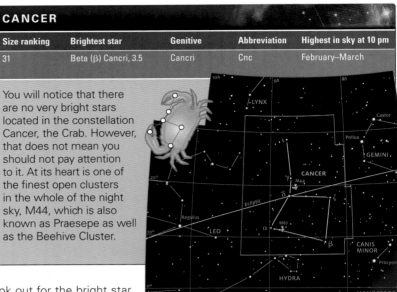

CANCER				
Size ranking	Brightest star	Genitive	Abbreviation	Highest in sky at 10 pm
31	Beta (β) Cancri, 3.5	Cancri	Cnc	February–March

You will notice that there are no very bright stars located in the constellation Cancer, the Crab. However, that does not mean you should not pay attention to it. At its heart is one of the finest open clusters in the whole of the night sky, M44, which is also known as Praesepe as well as the Beehive Cluster.

look out for the bright star Sirius shining away in the constellation Canis Major near the southwestern horizon.

SIGHTS OF INTEREST

If you are observing with a small telescope this month be sure to look out for the magnificent spiral galaxy M81 in the northern part of the constellation Ursa Major. On a clear March evening away from streetlights and other sources of light pollution M81 can be spotted through a pair of binoculars. Further south, look out for the well-placed Beehive Cluster, or M44, in the constellation Cancer.

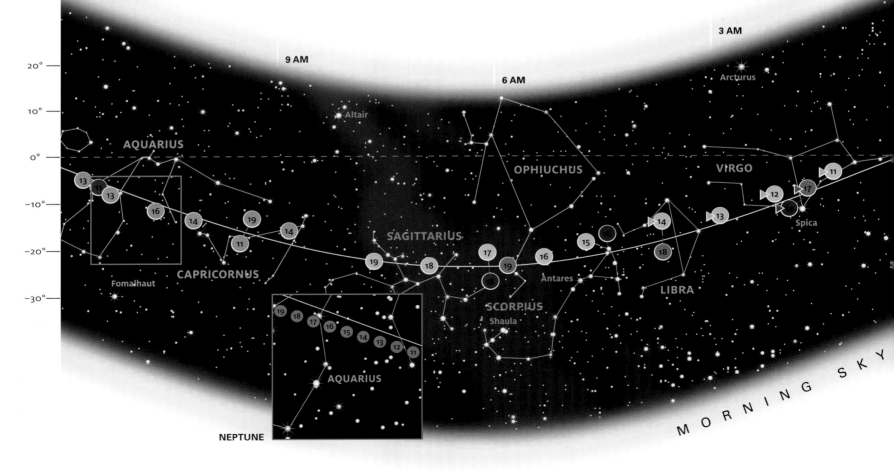

SOUTHERN LATITUDES

THE STARS

If you are observing the night sky from southern latitudes, your eyes will undoubtedly be drawn to the constellations sitting in the southeast, centred around Crux and Centaurus. From mid-latitudes, you can see Alphard, the brightest star in the constellation Hydra, sitting almost right above you.

Meanwhile Spica, the brightest star in Virgo, twinkles away in the east, with the blazing Canopus in Carina taking centre stage in the southwest sky. The constellation Orion is gradually sinking out of view, so make the most of it before it disappears. Leo is very much still on show and can be found sitting high in the northern part of the sky. Not far away from Leo, a little lower in the northwest, are the brightest stars of the constellation Gemini, Castor and Pollux.

SIGHTS OF INTEREST

You can see a great deal with a good pair of binoculars this month, including a lovely open cluster known as the Southern Pleiades,

The False Cross

Formed by four stars in the constellations Carina and Vela, the False Cross asterism resembles the constellation Crux, the Southern Cross, but is somewhat larger.

IC 2602. Its brightest member, the star Theta (θ) Carinae, can actually be spotted with the naked eye. If you turn binoculars on the cluster, you can see about 24 other sparkling stars. Around four degrees north of the Southern Pleiades is a glowing region NGC 3372, visible to the naked eye. Also known as the Carina Nebula, it is beautiful when observed through a small telescope.

VELA

Size ranking	Brightest star	Genitive	Abbreviation	Highest in sky at 10 pm
32	Gamma Velorum, 1.8	Velorum	Vel	February–April

The constellation Vela, the Sails, was once part of a larger constellation called Argo Navis, the Ship. Vela can be found in the night sky near the other parts of the ship, notably the constellations Carina, the Keel, and Puppis, the Stern.

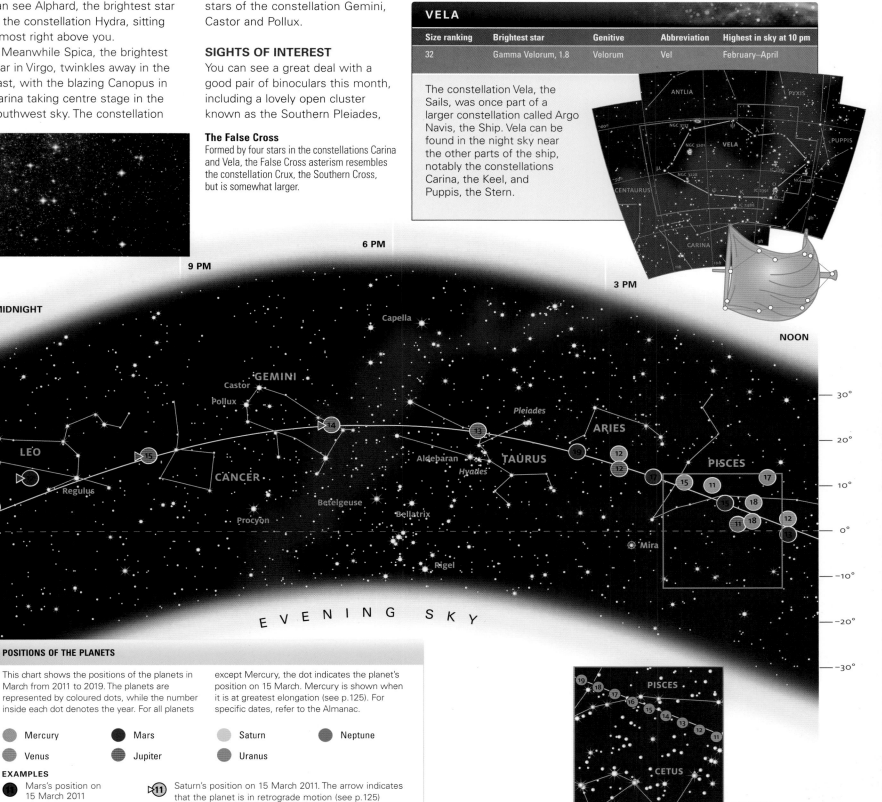

POSITIONS OF THE PLANETS

This chart shows the positions of the planets in March from 2011 to 2019. The planets are represented by coloured dots, while the number inside each dot denotes the year. For all planets except Mercury, the dot indicates the planet's position on 15 March. Mercury is shown when it is at greatest elongation (see p.125). For specific dates, refer to the Almanac.

- Mercury
- Venus
- Mars
- Jupiter
- Saturn
- Uranus
- Neptune

EXAMPLES

Mars's position on 15 March 2011

Saturn's position on 15 March 2011. The arrow indicates that the planet is in retrograde motion (see p.125)

MARCH
NORTHERN LATITUDES

OBSERVATION TIMES		
Date	Standard time	Daylight-saving time
15 February	Midnight	1 am
1 March	11 pm	Midnight
15 March	10 pm	11 pm
1 April	9 pm	10 pm
15 April	8 pm	9 pm

LOOKING NORTH

This month, look out for three beautiful star clusters in Taurus in the southwest: the Pleiades, M45, and the Hyades (see p.23). The Pleiades is arguably the finest open cluster in the northern skies. Although visible to the naked eye, it looks wonderful through all sorts of equipment – from binoculars to small, high-quality refractor telescopes. Other objects to spot in the vicinity are the open clusters NGC 1664 and NGC 1857 in Auriga, which are worth a look with a small telescope.

The Pleiades
Also known as the Seven Sisters, the Pleiades is a beautiful star cluster visible to the naked eye. It is a much loved target of both astrophotographers and astronomers observing with binoculars.

LOOKING SOUTH

Coma Berenices, Berenice's Hair, is a constellation located between Leo and Boötes in the west, and binoculars or a small telescope will reveal the scattered open star cluster known as Melotte 111, which lies within it. With a magnitude of 2.7, Melotte 111 is visible to the naked eye from a dark sky site.

Other objects to look out for from northern skies include the three galaxies in Leo: M65, M66, and NGC 3628, as well as the double star Algieba, Gamma (γ) Leonis.

Melotte 111
Also known as the Coma Star Cluster, Melotte 111 is an open cluster containing around 45 separate stars. It is a fine sight through a small refractor telescope or a pair of binoculars.

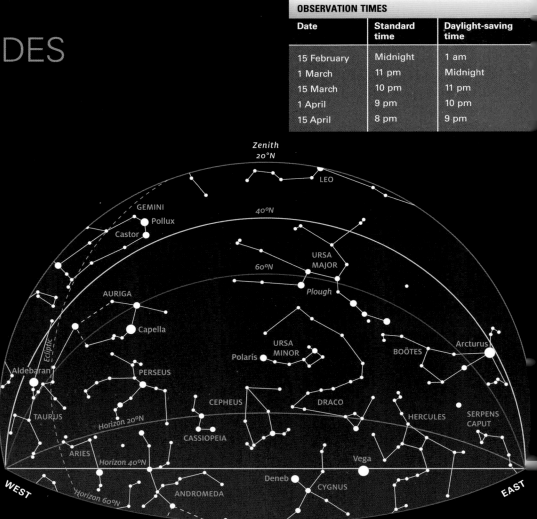

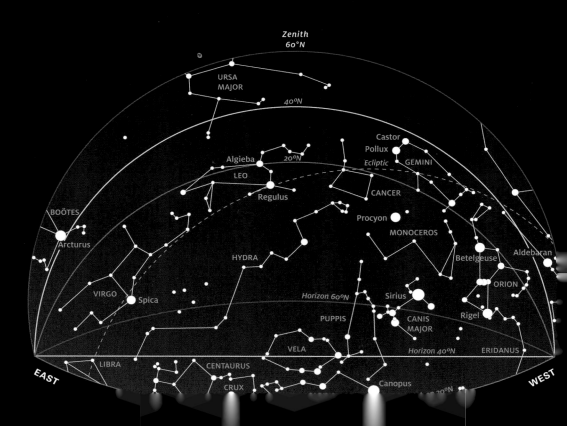

MARCH
SOUTHERN LATITUDES

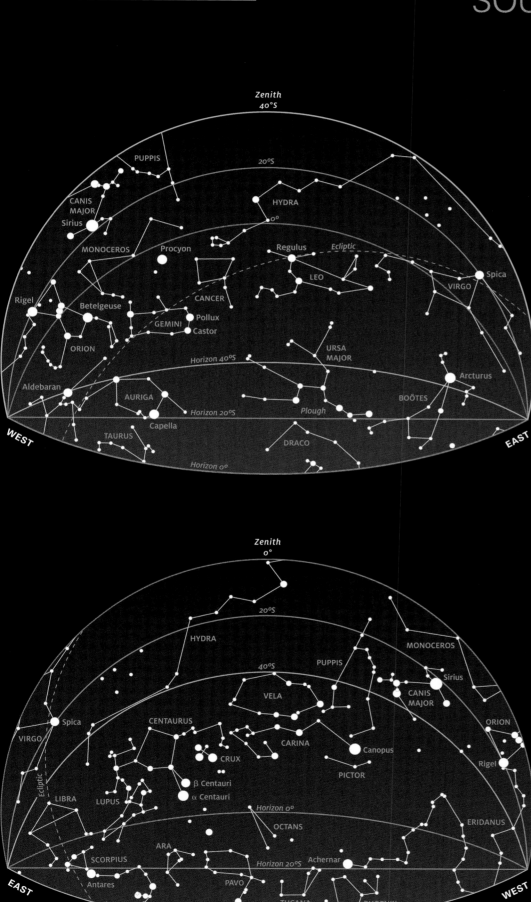

STAR MAGNITUDES

● -1   ● 0   ● 1   ● 2   • 3 and above

LOOKING NORTH

The galaxy M104 is an interesting target for deep-sky observers using a telescope. It sits in Virgo, the Virgin, which lies to the east in the southern skies. Also known as the Sombrero Galaxy, M104 is famous for a distinctive lane of dust that sits across its faintly glowing form. A relatively large telescope can show the dark lane clearly, but even a modest telescope reveals the galaxy's elliptical shape. Also in Virgo is the spiral galaxy M61; however, its low magnitude (9.7) makes it difficult to spot.

The Sombrero Galaxy
M104 is a good target if you have a large-aperture telescope. This stunning image from the Hubble Space Telescope shows the prominent dust lane in great detail.

LOOKING SOUTH

The globular cluster Omega (ω) Centauri is a must-see object for astronomers. A large telescope shows many of its stars, while binoculars show the cluster as a bright patch of light. In the west, the open cluster NGC 4755, the Jewel Box, in Crux, is a nice target for small telescopes and binoculars. Another western target is NGC 3372, or the Eta (η) Carinae Nebula a little further in Carina. NGC 3372 is visible to the naked eye against the Milky Way, with a dark lane of dust running through it.

Omega (ω) Centauri
Lying 17,000 light-years away, Omega (ω) Centauri or NGC 5139 is the largest globular cluster in the Milky Way. At magnitude 3.7, it is visible to the naked eye in the constellation Centaurus.

MARCH | NORTHERN LATITUDES

LOOKING NORTH

STAR MAGNITUDES

☀	-1
☀	0
☀	1
✷	2
·	3
·	4
·	5
⊙	Variable star

DEEP-SKY OBJECTS

🌀	Galaxy	✿	Globular cluster
✾	Open cluster	✵	Diffuse nebula
◉	Planetary nebula		

POINTS OF REFERENCE

Horizons	60°N	40°N	20°N
Zeniths	60°N	40°N	20°N
	Ecliptic	

OBSERVATION TIMES

Date	Standard time	Daylight-saving time
15 February	Midnight	1 am
1 March	11 pm	Midnight
15 March	10 pm	11 pm
1 April	9 pm	10 pm
15 April	8 pm	9 pm

MARCH | NORTHERN LATITUDES

LOOKING SOUTH

STAR MOTION

North

South

STAR MAGNITUDES

-1 · 0 · 1 · 2 · 3 · 4 · 5 ⊙ Variable star

DEEP-SKY OBJECTS

🌀 Galaxy ● Globular cluster ✳ Open cluster ◯ Diffuse nebula ◉ Planetary nebula

POINTS OF REFERENCE

Horizons | 60°N | 40°N | 20°N Zeniths + 60°N + 40°N + 20°N | Ecliptic

MARCH | SOUTHERN LATITUDES

LOOKING NORTH

STAR MAGNITUDES

★	-1
★	0
★	1
★	2
•	3
•	4
•	5
⊙	Variable star

DEEP-SKY OBJECTS

🌀	Galaxy
⚛	Globular cluster
✦	Open cluster
❀	Diffuse nebula
◉	Planetary nebula

POINTS OF REFERENCE

Horizons	0°	20°S	40°S
Zeniths	+ 0°	+ 20°S	+ 40°S
			Ecliptic

OBSERVATION TIMES

Date	Standard time	Daylight-saving time
15 February	Midnight	1 am
1 March	11 pm	Midnight
15 March	10 pm	11 pm
1 April	9 pm	10 pm
15 April	8 pm	9 pm

MARCH | SOUTHERN LATITUDES

LOOKING SOUTH

WEST

SOUTHWEST

SOUTH

SOUTHEAST

EAST

WEST

STAR MOTION

North

South

STAR MAGNITUDES

-1 · 0 · 1 · 2 · 3 · 4 · 5 ⊙ Variable star

DEEP-SKY OBJECTS

🌀 Galaxy ⬡ Globular cluster ✳ Open cluster ◌ Diffuse nebula ⬭ Planetary nebula

POINTS OF REFERENCE

Horizons | 0° | 20°S | 40°S Zeniths | 0° | 20°S | 40°S | Ecliptic

ERIDANUS
LEPUS
CANIS MAJOR
Sirius
M41
Adhara
FORNAX
CAELUM
COLUMBA
PICTOR
DORADO
HOROLOGIUM
PHOENIX
Achernar
RETICULUM
HYDRUS
MENSA
VOLANS
LMC
SMC
NGC 104
TUCANA
PYXIS
PUPPIS
M93
Canopus
CARINA
VELA
ANTLIA
CHAMAELEON
OCTANS
INDUS
NGC 3372
MUSCA
APUS
PAVO
HYDRA
CENTAURUS
Gacrux
Becrux
Acrux
NGC 4755
CRUX
β Centauri
α Centauri
CIRCINUS
TRIANGULUM AUSTRALE
NGC 5139
ARA
TELESCOPIUM
CORVUS
M83
NORMA
LUPUS
VIRGO
LIBRA
SCORPIUS
Shaula
M4
M6
M7
M80
Antares

APRIL

Although the nights in the northern hemisphere are getting shorter, there is still plenty of time to do some serious sky gazing. In the southern hemisphere there is plenty to see, including the magnificent arc of the Milky Way sweeping across the sky.

NORTHERN LATITUDES

THE STARS

The magnitude -0.1 star Arcturus in Boötes sits in the west this month, and should be one of your first signposts. Find it by following the curve of Ursa Major's handle, away from its "bowl". Along this curve, past Arcturus, you will eventually come across the bright star Spica in Virgo. Not far away from Virgo is Leo, and below these two constellations is a relatively empty patch of sky containing the long constellation Hydra, the Water Snake.

SIGHTS OF INTEREST

If you are an astronomer using binoculars the Coma Star Cluster in the constellation Coma Berenices is a beautiful object to observe on a clear April night. Look out for the

The Plough
The famous asterism known as the "Plough" sits high in the sky this month. It is part of the constellation Ursa Major, the Great Bear, and is sometimes referred to as the Big Dipper.

spiral galaxy M81 in Ursa Major with a small telescope. With a large telescope, you can seek out the Virgo Cluster containing many faint but interesting galaxies.

METEOR SHOWER

The Lyrid meteor shower is best seen from northern latitudes and reaches its peak around 21–22 April. The best time to view it is around dawn, when the bright star Vega, in Lyra, is highest in the sky. Although this shower does not create many meteors, they can be quite bright and fast; you can expect to see around 10 meteors over the course of an hour.

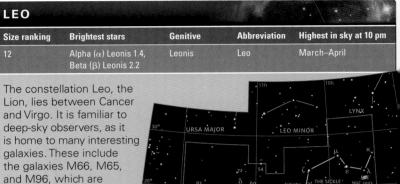

LEO				
Size ranking	Brightest stars	Genitive	Abbreviation	Highest in sky at 10 pm
12	Alpha (α) Leonis 1.4, Beta (β) Leonis 2.2	Leonis	Leo	March–April

The constellation Leo, the Lion, lies between Cancer and Virgo. It is familiar to deep-sky observers, as it is home to many interesting galaxies. These include the galaxies M66, M65, and M96, which are excellent targets for an amateur telescope. The constellation's brightest star is Regulus, Alpha (α) Leonis, which sits at the bottom of the famous backwards-question-mark-like "Sickle" asterism.

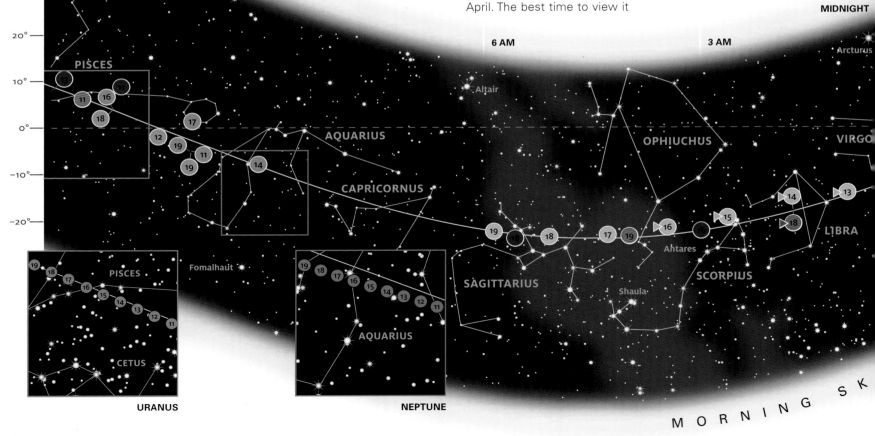

URANUS

NEPTUNE

MORNING SK

SOUTHERN LATITUDES

THE STARS

As the skies get dark you will notice that the constellations scattered along the arc of the Milky Way come into view. In the Southern skies, Crux, also known as the Southern Cross, and Centaurus, with the bright stars Rigil Kentaurus and Hadar, or Alpha (α) and Beta (β) Centauri are on show.

In the southeast, do not miss the bright star Antares in Scorpius. On the opposite side of the sky you will see Canopus in Carina. There is relatively little going on in the sky overhead, as this area is dominated by the long and winding constellation Hydra. However, you will find the bright star Spica in Virgo high in the east.

SIGHTS OF INTEREST

This time of the year is perfect for observing the dark nebula known as the Coalsack. You will find it nestled within the rich, bright star fields of the Milky Way. It is easily visible to the naked eye, sitting right next to Crux, or the Southern Cross. The Coalsack appears dark because it stops the light from the stars behind it from getting to our eyes.

Not far from the Coalsack you will find the wonderful open cluster NGC 4755, the Jewel Box Cluster. It looks like a hazy star to the naked eye, but binoculars or a small telescope will reveal its individual stars. If you are observing with binoculars, make sure you do not miss two marvellous sights in the nearby constellation Carina – IC 2602, or the Southern Pleiades, and NGC 3372, the Carina Nebula. The real star of the show is the stunning globular cluster NGC 5139, Omega (ω) Centauri, in the constellation Centaurus. A small telescope reveals many of its millions of stars.

The Coalsack
This dark nebula can be seen near Crux, the Southern Cross, with the naked eye. It is a vast dust cloud that blocks the light from the stars behind it.

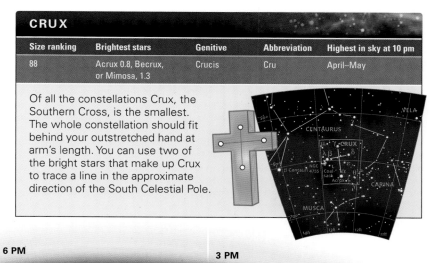

CRUX

Size ranking	Brightest stars	Genitive	Abbreviation	Highest in sky at 10 pm
88	Acrux 0.8, Becrux, or Mimosa, 1.3	Crucis	Cru	April–May

Of all the constellations Crux, the Southern Cross, is the smallest. The whole constellation should fit behind your outstretched hand at arm's length. You can use two of the bright stars that make up Crux to trace a line in the approximate direction of the South Celestial Pole.

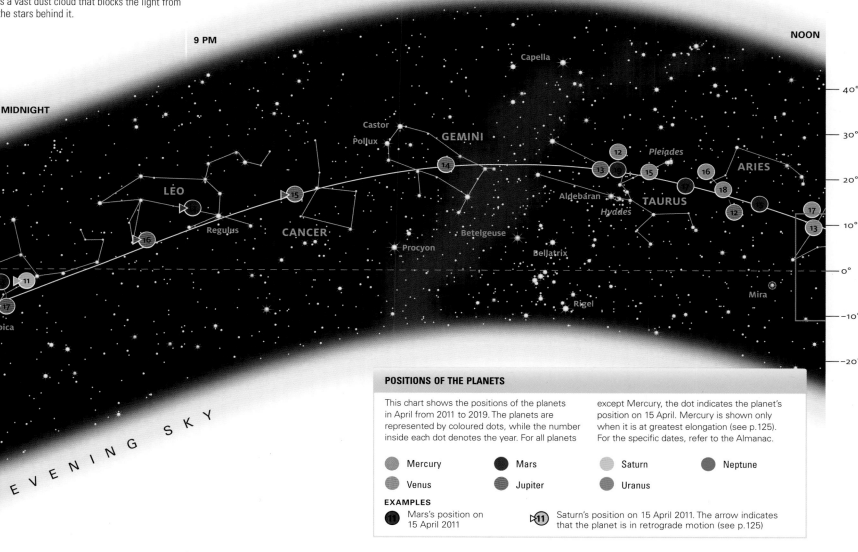

POSITIONS OF THE PLANETS

This chart shows the positions of the planets in April from 2011 to 2019. The planets are represented by coloured dots, while the number inside each dot denotes the year. For all planets except Mercury, the dot indicates the planet's position on 15 April. Mercury is shown only when it is at greatest elongation (see p.125). For the specific dates, refer to the Almanac.

- Mercury
- Venus
- Mars
- Jupiter
- Saturn
- Uranus
- Neptune

EXAMPLES

Mars's position on 15 April 2011

Saturn's position on 15 April 2011. The arrow indicates that the planet is in retrograde motion (see p.125)

APRIL
NORTHERN LATITUDES

OBSERVATION TIMES		
Date	Standard time	Daylight-saving time
15 March	Midnight	1 am
1 April	11 pm	Midnight
15 April	10 pm	11 pm
1 May	9 pm	10 pm
15 May	8 pm	9 pm

LOOKING NORTH

Northern skies in April contain the open cluster NGC 188 in the constellation Cepheus. This cluster sits just south of the bright star Polaris, Alpha (α) Ursae Minoris, and is a nice target for a large telescope.

For observers using binoculars, the open clusters M36, M37, and M38 in the constellation Auriga are still on show, as is the Double Cluster (see p.22) and the double star Mizar and Alcor. Also look out for the spiral galaxy M81 in the constellation Ursa Major.

M36 in Auriga
The open cluster M36 is a truly wonderful sight when seen through a small telescope. It sits in the middle of Messier's three famous open clusters in Auriga.

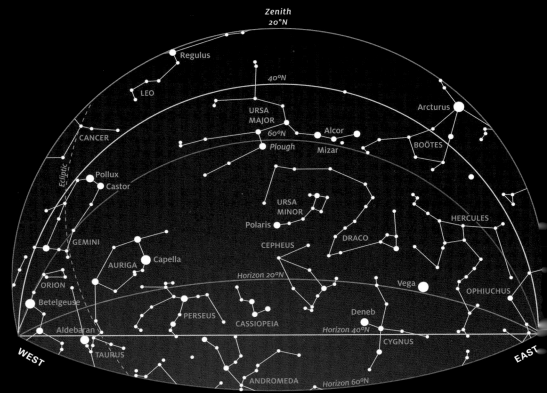

LOOKING SOUTH

There is a lot to see in April for deep-sky and galaxy enthusiasts looking south. Virgo has galaxies which can be seen through a telescope, such as M84, M86, and M87. The globular cluster M3 in Canes Venatici, the Hunting Dogs, makes a good small-telescope target. Canes Venatici can be located between Ursa Major and Boötes high in the northeast of the April skies. A small telescope will help show the lovely open cluster M48 in Hydra, the Water Snake. M48 is a loose collection of 80 stars southwest of Hydra's head.

The Virgo Galaxy Cluster
Lying in the constellation Virgo, this cluster is thought to consist of an incredible 2,000 individual galaxies. Several of its brightest members can be observed using amateur equipment.

APRIL
SOUTHERN LATITUDES

STAR MAGNITUDES

●-1 ●0 ●1 ●2 •3 and above

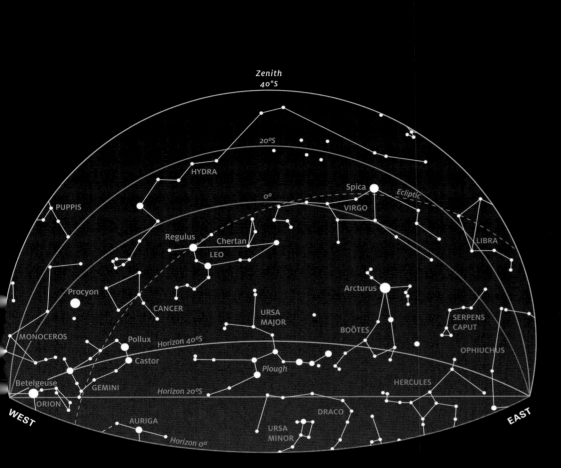

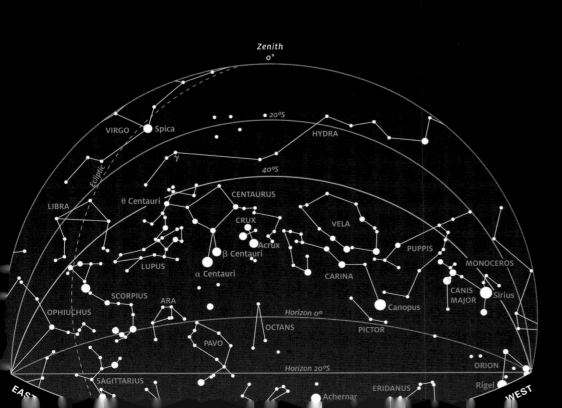

LOOKING **NORTH**

In the west, Leo plays host to many interesting galaxies that can be seen with relatively modest amateur equipment. The spiral galaxies M65 and M66 can be spotted with a small telescope, while M96, M95, and M105 are excellent targets for large-aperture telescopes. They are found clustered around a region at roughly the halfway point between the stars Chertan, or Theta (θ) Leonis, and Regulus, or Alpha (α) Leonis. M65 and M66 appear as grey smudges of light through a small telescope

M66 in Leo

The magnitude 8.9 spiral galaxy M66 appears in the bottom right of this image as part of the famous trio of galaxies known as the Leo Triplet, which also includes M65 and NGC 3628

LOOKING **SOUTH**

By far the most impressive sight from southern latitudes is the great arc of the Milky Way galaxy. It stretches all the way from the magnificent region in the east around the constellations Scorpius and Sagittarius through Crux, Carina, and Vela down to Puppis and Canis Major. The bright, magnitude 0.8 star Acrux, or Alpha (α) Crucis, in Crux is a multiple star, which can be resolved by a small telescope. The galaxy M83 in Hydra is an excellent large-aperture telescope object

M83 in Hydra

The spiral galaxy M83 sits in the constellation Hydra. It can be found in the night sky sitting between the stars Menkent, or

STAR MAGNITUDES

DEEP-SKY OBJECTS

POINTS OF REFERENCE

APRIL | NORTHERN LATITUDES

LOOKING NORTH

OBSERVATION TIMES		
Date	**Standard time**	**Daylight-saving time**
15 March	Midnight	1 am
1 April	11 pm	Midnight
15 April	10 pm	11 pm
1 May	9 pm	10 pm
15 May	8 pm	9 pm

WEST

STAR MOTION

North

South

SOUTHWEST

GEMINI

CANIS MINOR

MONOCEROS

M50

CANIS MAJOR

Sirius

M41

Adhara

Procyon

M48

M93

Adhara

M67

CANCER

M44

PUPPIS

PYXIS

HYDRA

LEO MINOR

URSA MAJOR

LEO

Regulus

SEXTANS

ANTLIA

VELA

CARINA

CRATER

COMA BERENICES

M86

M87 M84

M64

CORVUS

M104

M53

M3

CRUX

Gacrux

BOÖTES

Arcturus

VIRGO

Spica

CENTAURUS

Becrux

Acrux

NGC 5139

M83

β Centauri

SOUTH

SERPENS CAPUT

M5

LIBRA

LUPUS

Ecliptic

SCORPIUS

OPHIUCHUS

M12

M10

M80

M4

Antares

SOUTHEAST

EAST

WEST

LOOKING SOUTH

APRIL | NORTHERN LATITUDES

APRIL | SOUTHERN LATITUDES

LOOKING NORTH

OBSERVATION TIMES		
Date	**Standard time**	**Daylight-saving time**
15 March	Midnight	1 am
1 April	11 pm	Midnight
15 April	10 pm	11 pm
1 May	9 pm	10 pm
15 May	8 pm	9 pm

APRIL | SOUTHERN LATITUDES

LOOKING SOUTH

STAR MOTION

North

South

POINTS OF REFERENCE

Horizons | 0° | 20°S | 40°S   Zeniths ┼ 0° ┼ 20°S ┼ 40°S | Ecliptic

DEEP-SKY OBJECTS

🌀 Galaxy ● Globular cluster ✳ Open cluster ❀ Diffuse nebula ⬮ Planetary nebula

STAR MAGNITUDES

✴ -1 ✦ 0 ✦ 1 • 2 · 3 · 4 · 5 ⊙ Variable star

Constellations and objects labeled

WEST
EAST
SOUTH
SOUTHWEST
SOUTHEAST

ORION
LEPUS
CANIS MAJOR
Sirius
M50
M41
Adhara
M46
M47
M93
CAELUM
COLUMBA
PUPPIS
PYXIS
Canopus
DORADO
ERIDANUS
HOROLOGIUM
RETICULUM
Achernar
CARINA
PICTOR
VOLANS
MENSA
LMC
ANTLIA
VELA
CHAMAELEON
HYDRUS
PHOENIX
SMC
NGC 104
OCTANS
TUCANA
HYDRA
CENTAURUS
MUSCA
Gacrux
Acrux
Becrux
CRUX
β Centauri
APUS
CORVUS
NGC 5139
α Centauri
CIRCINUS
TRIANGULUM AUSTRALE
PAVO
M83
LUPUS
NORMA
ARA
INDUS
LIBRA
SCORPIUS
TELESCOPIUM
M80
M4
Antares
M62
Shaula
M19
M6
M7
CORONA AUSTRALIS
SAGITTARIUS
M69
M54
M9
M8
M28
M10
M23
M21
M24
OPHIUCHUS

MAY

In the southern hemisphere this month you will be treated to the richness of the constellations Centaurus, Scorpius, and Sagittarius. If you are observing from the northern hemisphere, you will see more subdued constellations, such as Hercules and Virgo.

NORTHERN LATITUDES

THE STARS

If you have a small telescope, point it at the middle star in the "handle" of the Plough asterism in Ursa Major. This is the star Mizar, which has a companion star, Alcor, that can be glimpsed with the naked eye. A closer inspection of Mizar with a telescope shows that it is made up of a pair of stars. Once again, follow the Plough's curving handle to find your way to Arcturus in the constellation Boötes. To its south, you will see the bright star Spica in Virgo. This month the bright blue-white star Vega rises in the east in the constellation Lyra, the Lyre – a celestial sign that summer is on its way. If you are observing the night sky from lower northerly latitudes, you can also glimpse the constellation Scorpius, the Scorpion, peeking over the southeastern horizon, led by the bright orange-red star Antares.

Finding the Pole star
You can use the stars Alpha (α) and Beta (β) Ursae Majoris (right) in the Plough asterism to find the location of the pole star, Polaris (centre top).

SIGHTS OF INTEREST

For galaxy enthusiasts, the May night sky offers two relatively bright targets. The first is the Whirlpool Galaxy, or M51, in Canes Venatici. Second is the spiral galaxy M101, which sits to the north of the Plough's handle.

METEOR SHOWER

The annual Eta Aquarid meteor shower peaks this month. As the radiant of the shower lies near the celestial equator, it is not a great sight from far northerly latitudes.

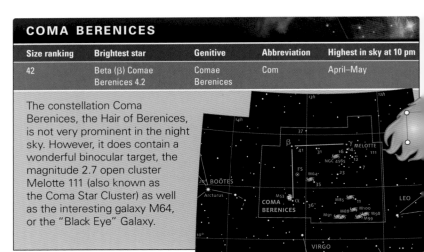

COMA BERENICES

Size ranking	Brightest star	Genitive	Abbreviation	Highest in sky at 10 pm
42	Beta (β) Comae Berenices 4.2	Comae Berenices	Com	April–May

The constellation Coma Berenices, the Hair of Berenices, is not very prominent in the night sky. However, it does contain a wonderful binocular target, the magnitude 2.7 open cluster Melotte 111 (also known as the Coma Star Cluster) as well as the interesting galaxy M64, or the "Black Eye" Galaxy.

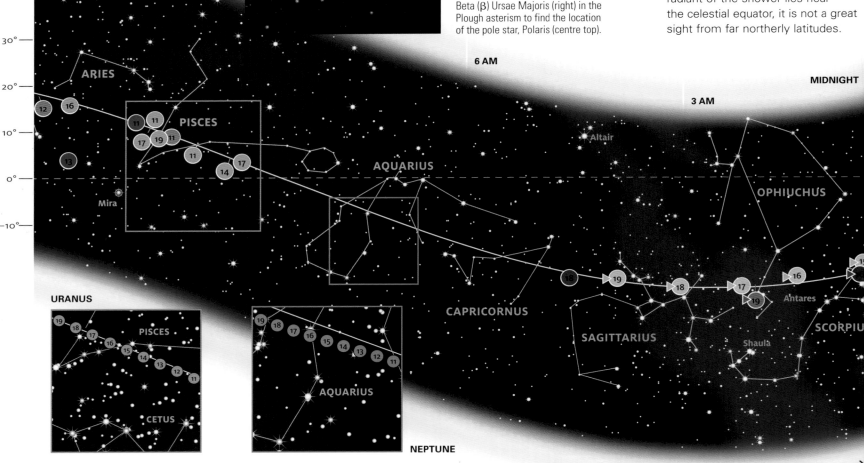

SOUTHERN LATITUDES

THE STARS

Sitting high in the southern sky is the prominent constellation Crux, the Southern Cross. If you have a small telescope, be sure to point it at Crux's brightest star Acrux, Alpha (α) Crucis, which is actually a double star made of two blue-white stars.

The Milky Way rises high in the south this month, with Sagittarius and Scorpius sitting in the east – a clue that winter is on its way in the southern hemisphere.

SIGHTS OF INTEREST

Whether you are observing with the naked eye or a telescope, it is the globular cluster NGC 5139, Omega (ω) Centauri, you will want to set your sights on this month. You can see it easily as a hazy star with the naked eye, whereas a telescope reveals many of its millions of stars in a tight ball. If you have a large-aperture telescope, turn it towards the fine spiral galaxy M83 that sits in the constellation Hydra.

METEOR SHOWER

Caused by the dust left over from Comet Halley entering our atmosphere and vaporizing, the Eta (η) Aquarid meteor shower peaks every year around 5–6 May. You can expect to see roughly 30 meteors an hour. The meteors appear to be coming from a point near the star Eta (η) Aquarii, in Aquarius, and tend to be quite fast moving. The further south you are, the better view of the shower you will get.

Spectacular star fields
Looking into the southern night sky in May you cannot miss the bright stars Alpha (α) and Beta (β) Centauri (left). Nearby you will find the constellation Crux (right) and the Coalsack Nebula.

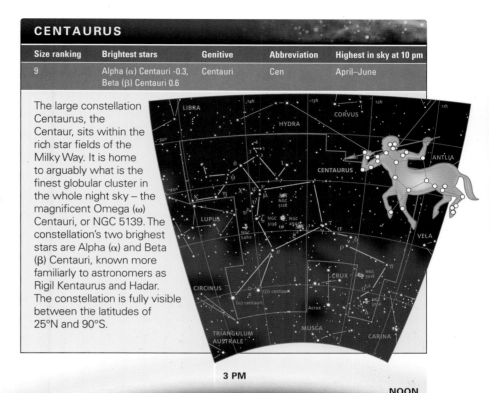

CENTAURUS

Size ranking	Brightest stars	Genitive	Abbreviation	Highest in sky at 10 pm
9	Alpha (α) Centauri -0.3, Beta (β) Centauri 0.6	Centauri	Cen	April–June

The large constellation Centaurus, the Centaur, sits within the rich star fields of the Milky Way. It is home to arguably what is the finest globular cluster in the whole night sky – the magnificent Omega (ω) Centauri, or NGC 5139. The constellation's two brightest stars are Alpha (α) and Beta (β) Centauri, known more familiarly to astronomers as Rigil Kentaurus and Hadar. The constellation is fully visible between the latitudes of 25°N and 90°S.

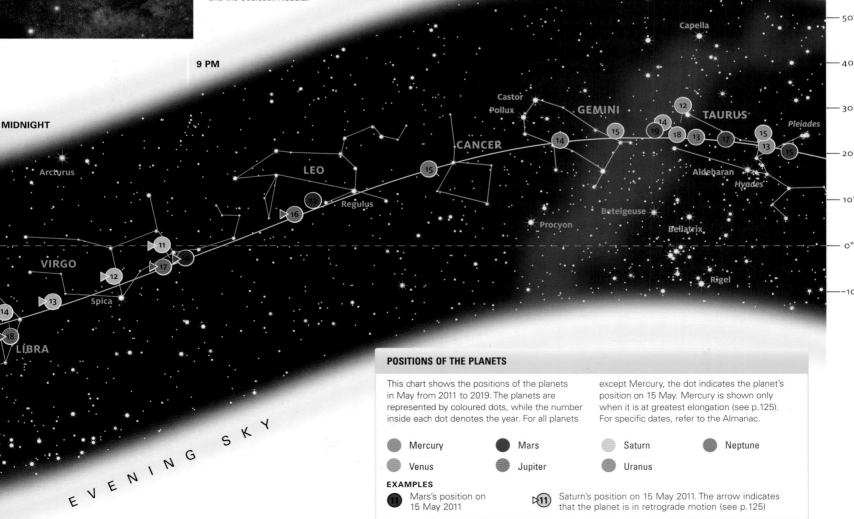

POSITIONS OF THE PLANETS

This chart shows the positions of the planets in May from 2011 to 2019. The planets are represented by coloured dots, while the number inside each dot denotes the year. For all planets except Mercury, the dot indicates the planet's position on 15 May. Mercury is shown only when it is at greatest elongation (see p.125). For specific dates, refer to the Almanac.

- Mercury
- Venus
- Mars
- Jupiter
- Saturn
- Uranus
- Neptune

EXAMPLES

🔵11 Mars's position on 15 May 2011

▷11 Saturn's position on 15 May 2011. The arrow indicates that the planet is in retrograde motion (see p.125)

MAY
NORTHERN LATITUDES

OBSERVATION TIMES

Date	Standard time	Daylight-saving time
15 April	Midnight	1 am
1 May	11 pm	Midnight
15 May	10 pm	11 pm
1 June	9 pm	10 pm
15 June	8 pm	9 pm

LOOKING **NORTH**

The globular cluster M13 in Hercules is a prominent feature in May. It is located roughly a third of the way along a line between the stars Eta (η) and Zeta (ζ) Herculis. Although M13 can be seen with binoculars, a telescope shows it more clearly. Larger aperture telescopes can show a multitude of the cluster's stars.

Another object to look out for is the planetary nebula NGC 6543 in Draco, which is best suited to larger telescopes.

NGC 6543
This magnitude 8.1 nebula, also known as the Cat's Eye Nebula, is best suited for larger telescopes and appears as a bluish disc. It is 3,600 light-years away from Earth.

LOOKING **SOUTH**

There are several notable globular clusters worth observing in May's night sky. M10 in the east, at the heart of the constellation Ophiuchus, is visible with binoculars. Just northwest of M10 but also in Ophiuchus lies another cluster, M12, which is a fine sight through a small telescope.

Moving up a little, the globular cluster M5 appears over the border in the southern part of Serpens Caput, the Snake's Head. Kappa (κ) Boötis is a double star in Boötes, visible through a small telescope.

M10 in Ophiuchus
The magnitude 6.6 globular cluster M10 lies 14,000 light-years from Earth, and is a little over 80 light-years in diameter. It is a wonderful target for a small amateur telescope.

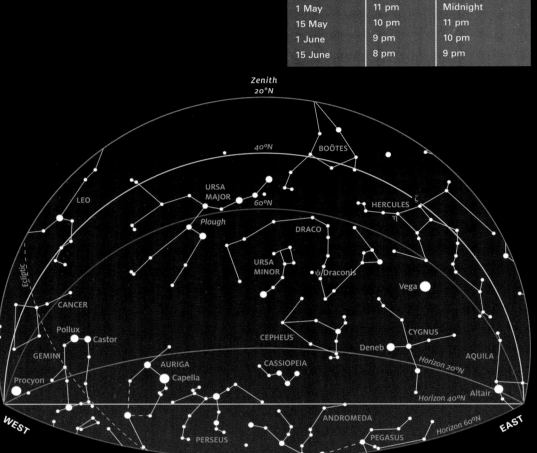

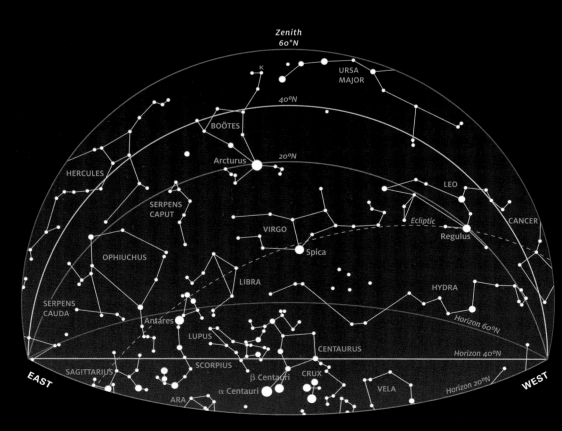

MAY
SOUTHERN LATITUDES

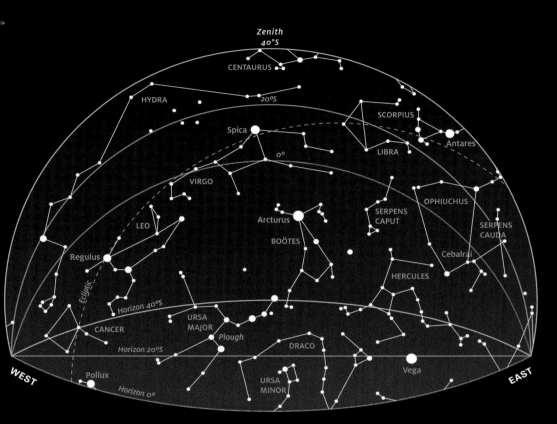

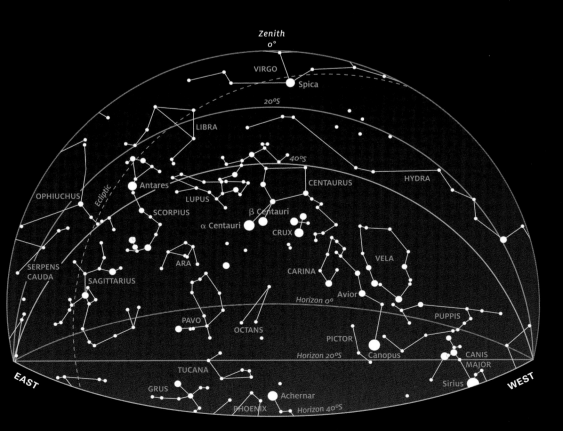

LOOKING **NORTH**

In the east, several star clusters are on view in the constellation Ophiuchus this month. Besides the globular clusters M10 and M12, there is the magnitude 4.6 open cluster NGC 6633, that is roughly the same size as the full Moon on the sky. Composed of 30 stars, this open cluster is a wonderful sight through a small telescope. Northwest of NGC 6633 is another large and scattered open cluster IC 4665, which lies close to the star Cebalrai, or Beta (β) Ophiuchi, and is easily visible with binoculars.

M12 in Ophiuchus
Discovered in 1764 by Charles Messier, M12 is a prime target for a small telescope. It is estimated to be between 16,000 and 18,000 light-years away from Earth.

LOOKING **SOUTH**

The beautiful section of the Milky Way around Crux, the Southern Cross, including the dark nebula known as the Coalsack, takes centre stage in the south in May. There are several fine open clusters on view in Carina, many nestled among the rich star fields of the Milky Way. NGC 3532 can be seen with the naked eye, but binoculars reveal its many twinkling stars well. The magnitude 4.2 cluster NGC 3114 is an interesting target for a small telescope, while NGC 2516 is a good target for binoculars.

NGC 2516
The magnitude 3.8 open cluster NGC 2516 sits roughly 3.5 degrees away from the star Avior, Epsilon (ε) Carinae. It contains roughly 100 stars and can be viewed with just a pair of binoculars.

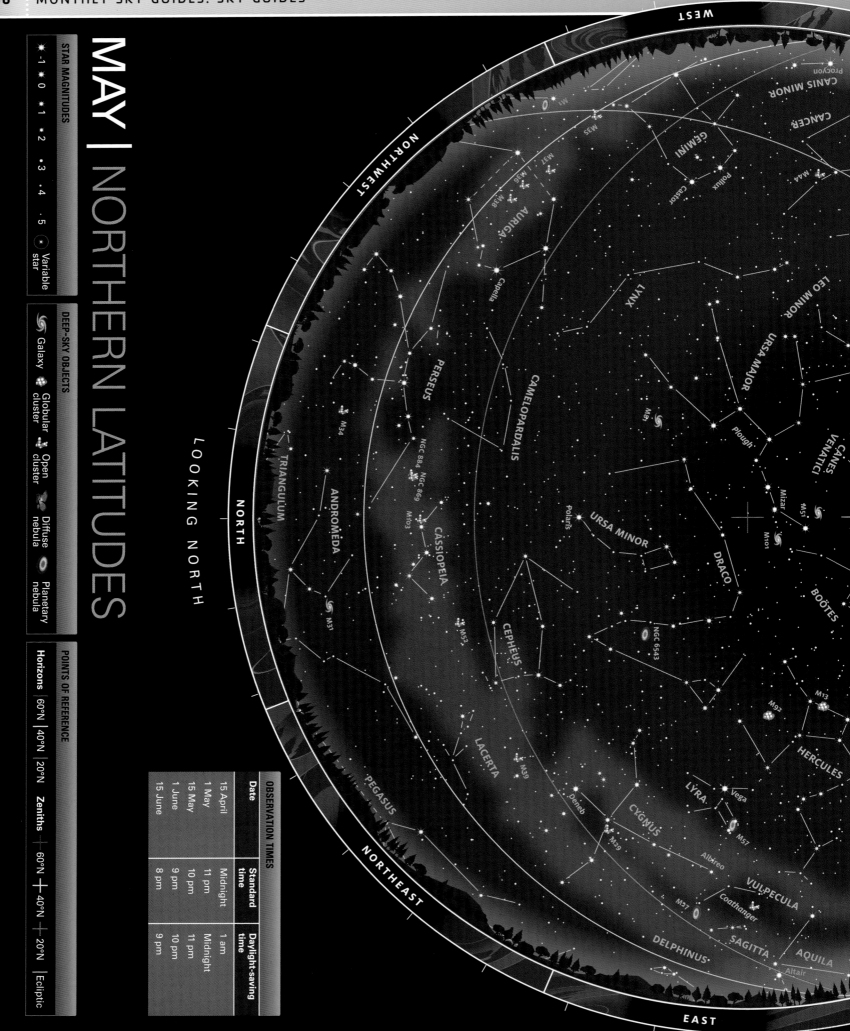

MAY | NORTHERN LATITUDES

LOOKING NORTH

STAR MAGNITUDES

✦	-1
✦	0
✶	1
∗	2
·	3
·	4
·	5
⊙	Variable star

DEEP-SKY OBJECTS

🌀	Galaxy
⊛	Globular cluster
❀	Open cluster
❁	Diffuse nebula
◉	Planetary nebula

POINTS OF REFERENCE

Horizons | 60°N | 40°N | 20°N

Zeniths | 60°N | 40°N | 20°N | Ecliptic

OBSERVATION TIMES

Date	Standard time	Daylight-saving time
15 April	Midnight	1 am
1 May	11 pm	Midnight
15 May	10 pm	11 pm
1 June	9 pm	10 pm
15 June	8 pm	9 pm

MAY | NORTHERN LATITUDES

LOOKING SOUTH

STAR MOTION

North

South

EAST

SOUTHEAST

SOUTH

SOUTHWEST

WEST

STAR MAGNITUDES

★	★	★	•	•	•	•	⊙
-1	0	1	2	3	4	5	Variable star

DEEP-SKY OBJECTS

🌀 Galaxy	● Globular cluster	✳ Open cluster	🌫 Diffuse nebula	⊙ Planetary nebula

POINTS OF REFERENCE

Horizons	60°N	40°N	20°N	Zeniths	60°N	40°N	20°N	Ecliptic

Constellations and objects

BOÖTES
CORONA BOREALIS
SERPENS CAPUT
HERCULES
OPHIUCHUS
SERPENS CAUDA
LEO
HYDRA
SEXTANS
VIRGO
COMA BERENICES
CRATER
CORVUS
ANTLIA
PYXIS
VELA
CENTAURUS
CRUX
LIBRA
SCORPIUS
LUPUS
NORMA
CIRCINUS
ARA

Regulus
Arcturus
Spica
Antares
Shaula
Regulus
Gacrux
Becrux
α Crux
α Centauri
β Centauri

M67
M48
M3
M53
M64
M87
M104
M83
M5
M80
M4
M12
M10
M19
M62
M14
M9
M6
M23
M21
M8
M16
M19
M24
M28
M17
M25
M22
M26
M11
Ecliptic

MAY | SOUTHERN LATITUDES

STAR MAGNITUDES

* -1
* 0
* 1
* 2
* 3
* 4
* 5
* Variable star

DEEP-SKY OBJECTS

* Galaxy
* Globular cluster
* Open cluster
* Diffuse nebula
* Planetary nebula

POINTS OF REFERENCE

Horizons | 0° | 20°S | 40°S

Zeniths | 0° | 20°S | 40°S | Ecliptic

LOOKING NORTH

OBSERVATION TIMES		
Date	Standard time	Daylight-saving time
15 April	Midnight	1 am
1 May	11 pm	Midnight
15 May	10 pm	11 pm
1 June	9 pm	10 pm
15 June	8 pm	9 pm

MAY | SOUTHERN LATITUDES

LOOKING SOUTH

STAR MOTION

North

South

POINTS OF REFERENCE

| Horizons | 0° | 20°S | 40°S | Zeniths | 0° | 20°S | 40°S | Ecliptic |

DEEP-SKY OBJECTS

Galaxy | Globular cluster | Open cluster | Diffuse nebula | Planetary nebula

STAR MAGNITUDES

-1 | 0 | 1 | 2 | 3 | 4 | 5 | Variable star

JUNE

It is summer in the northern hemisphere, and with lighter evenings the time for observation is reduced. For observers in the southern hemisphere, the dark skies offer a plethora of celestial sights to look out for, including the constellations in the Milky Way.

NORTHERN LATITUDES

THE STARS

Looking north Ursa Minor, the Little Bear, is clearly visible with Draco, the Dragon, wrapped around it. The tip of the bear's tail is marked by the pole star, Polaris.

If observing from a site with a clear southern horizon, you will be able to spot the constellation Scorpius. Look out for the unmistakable bright star Alpha (α) Scorpii, Antares, shining with an orange-red tint.

SIGHTS OF INTEREST

If you have a small telescope train it on M13, the finest globular cluster in the northern skies this month. It lies in Hercules, which is high in the sky at this time. Also look out for another globular cluster M5, which sits in the head of the constellation Serpens, the Snake. These star clusters are roughly magnitude 6 and can be spotted through binoculars. If you are a keen galaxy observer, use a telescope to reveal two well-known spiral galaxies, M51 and M101, sitting near the "Plough's" handle.

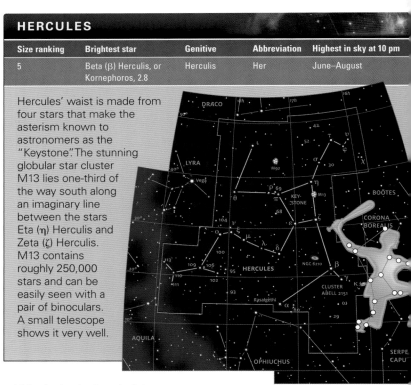

HERCULES

Size ranking	Brightest star	Genitive	Abbreviation	Highest in sky at 10 pm
5	Beta (β) Herculis, or Kornephoros, 2.8	Herculis	Her	June–August

Hercules' waist is made from four stars that make the asterism known to astronomers as the "Keystone". The stunning globular star cluster M13 lies one-third of the way south along an imaginary line between the stars Eta (η) Herculis and Zeta (ζ) Herculis. M13 contains roughly 250,000 stars and can be easily seen with a pair of binoculars. A small telescope shows it very well.

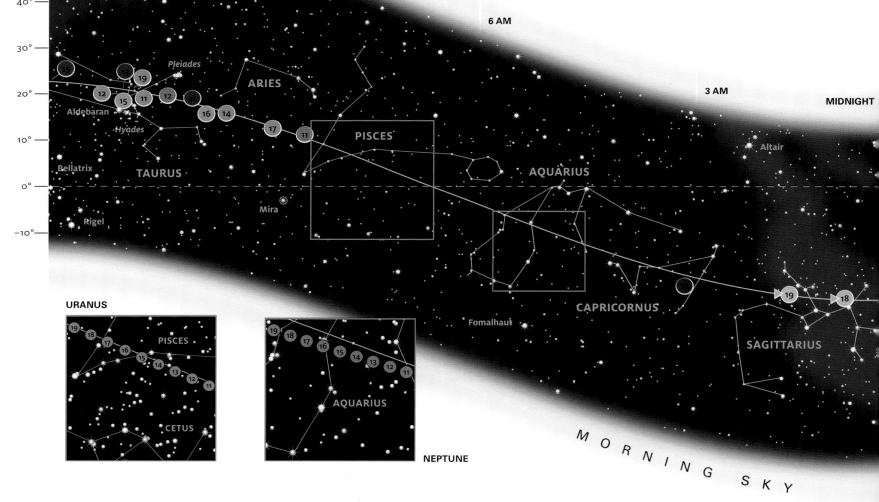

SOUTHERN LATITUDES

THE STARS

June is a wonderful time for night-sky observers in the southern hemisphere. The rich star fields of the Milky Way galaxy stretch right across the sky from the southwest to the northeast. Scattered among them are the sparkling constellations Centaurus, the Centaur; Crux, the Southern Cross; as well as Scorpius, the Scorpion; Carina, the Keel; and Sagittarius the Archer.

High in the south you will see the constellation Lupus, while the distinctive shape of Scorpius will help you get your bearings. Its brightest star is the orange-red Antares. Be sure to scan your eyes across the beautiful star fields in the constellation Sagittarius, especially if you are observing from a dark sky site. Looking north the constellations Boötes, Hercules, and Ophiuchus are on show.

SIGHTS OF INTEREST

There is no shortage of interesting objects to view from southern latitudes this month. A good place to start is the constellation Scorpius, which contains the stunning open clusters M6 and M7, both of which can be viewed with the naked eye. They sit not far from the Scorpion's tail and a pair of binoculars shows them very clearly. There is also the open star cluster NGC 6231 nearby, lying close to the star Zeta (ζ) Scorpii. The magnificent globular cluster Omega (ω) Centauri in the constellation Centaurus is still on show. It is breathtaking when seen with a large telescope.

Also in the south, not far away from Centaurus, in the constellation Crux, you can find the dark Coalsack Nebula. You can also view the spectacular Jewel Box Cluster and M83, a spiral galaxy in Hydra.

Sparkling Scorpius
The distinctive constellation Scorpius is rich in deep-sky objects to observe, including the marvellous open clusters M6 and M7, found near its tail (top left).

SCORPIUS

Size ranking	Brightest star	Genitive	Abbreviation	Highest in sky at 10 pm
33	Alpha (α) Scorpii, or Antares, 1.0	Scorpii	Sco	June–July

You cannot fail to spot the constellation Scorpius, which has one of the most recognizable patterns in the night sky. It is home to many excellent targets for an amateur telescope. However, to see the whole of the constellation in the night sky your location should be to the south of latitude 40 degrees north. Scorpius' brightest star is the orange-red Antares, Alpha (α)Scorpii. It is a supergiant star with a diameter about 800 times that of our own star, the Sun.

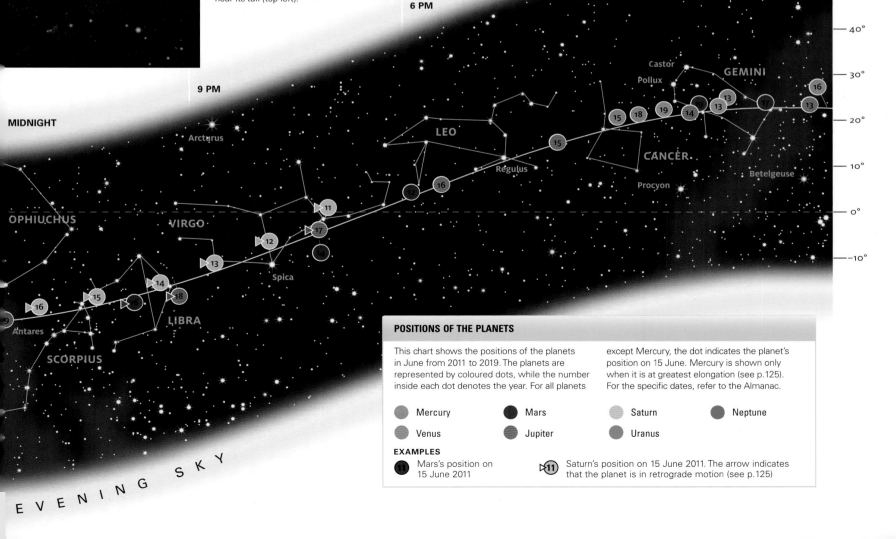

POSITIONS OF THE PLANETS

This chart shows the positions of the planets in June from 2011 to 2019. The planets are represented by coloured dots, while the number inside each dot denotes the year. For all planets except Mercury, the dot indicates the planet's position on 15 June. Mercury is shown only when it is at greatest elongation (see p.125). For the specific dates, refer to the Almanac.

- Mercury
- Venus
- Mars
- Jupiter
- Saturn
- Uranus
- Neptune

EXAMPLES

⬤11 Mars's position on 15 June 2011

▷11 Saturn's position on 15 June 2011. The arrow indicates that the planet is in retrograde motion (see p.125)

EVENING SKY

JUNE
NORTHERN LATITUDES

LOOKING NORTH

One of the finest double stars of the northern skies is on show in the east at this time of the year. Albireo, or Beta (β) Cygni, in Cygnus is a must-see for any beginner due to the ease with which the stars can be separated. It can be seen through a small telescope, with one of its stars shining gold and the other with a blue hue. The open cluster M39, also in Cygnus, is another good small-telescope target, as is the variable star Delta (δ) Cephei that varies between magnitude 3.5 and 4.4 every 5 days and 9 hours.

Albireo
A small telescope shows the striking colour difference of the two stars, set against the star fields of the Milky Way in Cygnus. The magnitude 3.1 and 5.1 stars lie 380 light-years away from Earth.

LOOKING SOUTH

The constellation Boötes, the Herdsman, is high in the sky in June. Its brightest star Arcturus, with a magnitude -0.04, is a red giant star that is an incredible 25 times larger than our Sun.

Also still on view, to the east of Boötes, is the globular cluster M13 in Hercules. Low down on the horizon is Scorpius, which contains some interesting objects, such as two open clusters, M6 and M7, and a globular cluster M4. Both M6 and M7 are visible to the naked eye and their individual stars can be seen through binoculars.

M6 in Scorpius
A magnitude 4.2 open cluster, M6 can be found sitting not far from the "sting" in the tail of Scorpius, just north of M7. It is also known as the Butterfly Cluster.

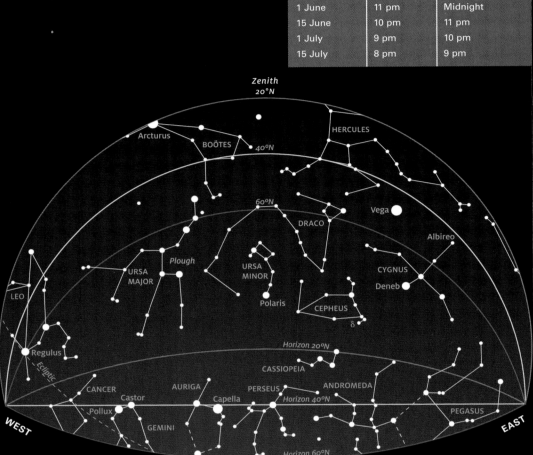

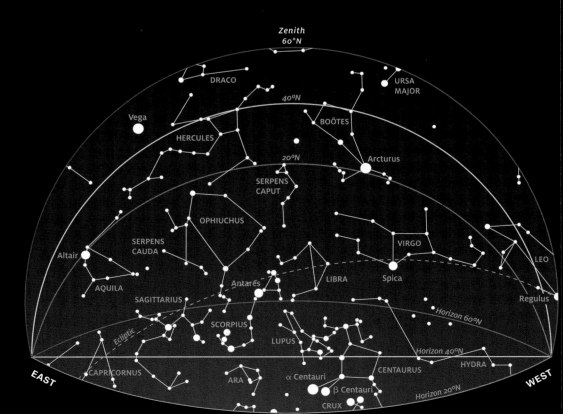

JUNE
SOUTHERN LATITUDES

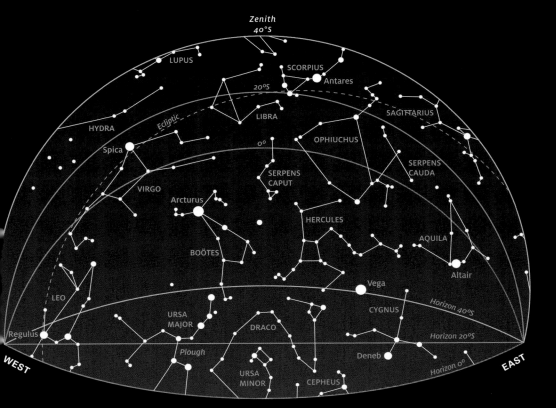

Zenith
40°S

LUPUS
SCORPIUS
Antares
20°S
Ecliptic
HYDRA
LIBRA
SAGITTARIUS
Spica
0°
OPHIUCHUS
SERPENS CAUDA
VIRGO
SERPENS CAPUT
Arcturus
HERCULES
BOÖTES
AQUILA
Altair
LEO
Vega
Horizon 40°S
URSA MAJOR
CYGNUS
Regulus
DRACO
Horizon 20°S
Plough
Deneb
WEST
URSA MINOR
CEPHEUS
Horizon 0°
EAST

LOOKING **NORTH**

While there may be more impressive sights in the southern part of the June sky, there is also much to see when looking north, such as the globular clusters M13 and M92 in the constellation Hercules. There are also plenty of interesting clusters to look at in Ophiuchus. Besides the two globular clusters M10 and M12, and the open cluster NGC 6633, be sure to look out for the magnitude 4.2 open cluster IC 4665. It is composed of a group of 30 stars and is a lovely sight through binoculars.

M13 in Hercules
The globular cluster M13 in Hercules is a spectacular sight in a large-aperture telescope. A large Dobsonian telescope, for example, will show it as a ball of thousands of stars.

LOOKING **SOUTH**

If you are looking south in the southern hemisphere, you will be met with a rich variety of objects visible with just the naked eye, or with binoculars or a small telescope. M22 in the constellation Sagittarius is an impressive magnitude 5.1 globular cluster, while the emission nebula, M8, is a fine target for binoculars. Meanwhile, Omega (ω) Centauri, arguably the finest globular cluster in the night sky, sits at the heart of Centaurus, the Centaur.

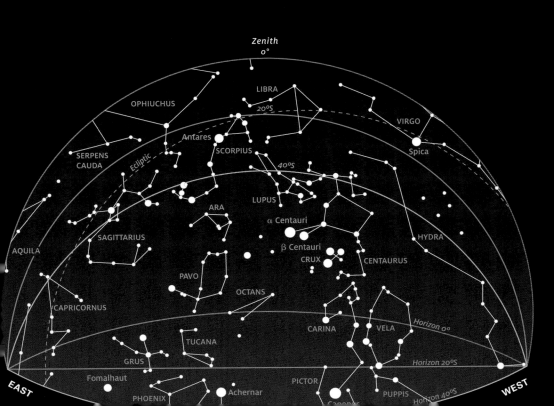

Zenith
0°

LIBRA
OPHIUCHUS
20°S
VIRGO
Antares
Ecliptic
SCORPIUS
Spica
SERPENS CAUDA
40°S
LUPUS
ARA
AQUILA
SAGITTARIUS
α Centauri
HYDRA
β Centauri
CRUX
CENTAURUS
PAVO
OCTANS
CAPRICORNUS
Horizon 0°
CARINA
VELA
TUCANA
Horizon 20°S
GRUS
Fomalhaut
PICTOR
PUPPIS
Horizon 40°S
EAST
PHOENIX
Achernar
WEST

M8 in Sagittarius
Also called the Lagoon Nebula, M8 can be seen through binoculars as a glowing patch. The view through a large telescope is mesmerizing, with several stars nestled in and around the nebula.

JUNE | NORTHERN LATITUDES

LOOKING NORTH

STAR MAGNITUDES

- ✴ -1
- ✴ 0
- ✴ 1
- ✴ 2
- • 3
- • 4
- • 5
- ⊙ Variable star

DEEP-SKY OBJECTS

- Galaxy
- Globular cluster
- Open cluster
- Diffuse nebula
- Planetary nebula

POINTS OF REFERENCE

Horizons	60°N	40°N	20°N
Zeniths	60°N	40°N	20°N
			Ecliptic

OBSERVATION TIMES		
Date	**Standard time**	**Daylight-saving time**
15 May	Midnight	1 am
1 June	11 pm	Midnight
15 June	10 pm	11 pm
1 July	9 pm	10 pm
15 July	8 pm	9 pm

Constellation and star labels on the chart: WEST, NORTHWEST, NORTH, NORTHEAST, EAST, GEMINI, CANCER, LEO, Regulus, LEO MINOR, Pollux, Castor, LYNX, URSA MAJOR, Plough, CANES VENATICI, M51, Mizar, M101, BOÖTES, AURIGA, Capella, M37, M36, M38, CAMELOPARDALIS, M81, Polaris, URSA MINOR, DRACO, M92, HERCULES, Vega, PERSEUS, NGC 884, NGC 869, M34, M103, CASSIOPEIA, CEPHEUS, M52, LYRA, M57, CYGNUS, Albireo, Deneb, M39, TRIANGULUM, M33, M39, VULPECULA, M27, LACERTA, ANDROMEDA, DELPHINUS, PEGASUS, EQUULEUS, M15

STAR MOTION

North

South

LOOKING SOUTH

JUNE | NORTHERN LATITUDES

SEXTANS

LEO

COMA BERENICES

CRATER

HYDRA

CORVUS

VIRGO

M87

M64

M53

M3

BOÖTES

Arcturus

CORONA BOREALIS

SERPENS CAPUT

M5

Spica

Ecliptic

M83

M104

NGC 5139

CENTAURUS

Gacrux

Becrux

β Centauri

α Centauri

LUPUS

LIBRA

CIRCINUS

NORMA

TRIANGULUM AUSTRALE

SCORPIUS

Antares

M4

M80

M19

M62

Shaula

M6

M7

ARA

TELESCOPIUM

CORONA AUSTRALIS

SAGITTARIUS

M55

M54

M69

M22

M28

M8

M24

M25

M21

M23

M20

M17

M16

M18

M26

M11

SCUTUM

M14

M10

M12

OPHIUCHUS

SERPENS CAUDA

HERCULES

M13

VULPECULA

SAGITTA

Altair

AQUILA

CAPRICORNUS

M9

SOUTHWEST

SOUTH

SOUTHEAST

EAST

POINTS OF REFERENCE

	Horizons	60°N	40°N	Zeniths	60°N	40°N	Ecliptic

40°N | 20°N   20°N

DEEP-SKY OBJECTS

Galaxy   Globular cluster   Open cluster   Diffuse nebula   Planetary nebula

STAR MAGNITUDES

-1   0   1   2   3   4   5   Variable star

JUNE | SOUTHERN LATITUDES

STAR MAGNITUDES

* -1
* 0
* 1
* 2
* 3
* 4
* 5
* Variable star

DEEP-SKY OBJECTS

* Galaxy
* Globular cluster
* Open cluster
* Diffuse nebula
* Planetary nebula

POINTS OF REFERENCE

Horizons | 0° | 20°S | 40°S
Zeniths | 0° | 20°S | 40°S | Ecliptic

LOOKING NORTH

OBSERVATION TIMES

Date	Standard time	Daylight-saving time
15 May	Midnight	1 am
1 June	11 pm	Midnight
15 June	10 pm	11 pm
1 July	9 pm	10 pm
15 July	8 pm	9 pm

JUNE | SOUTHERN LATITUDES

LOOKING SOUTH

STAR MOTION

North

South

POINTS OF REFERENCE

| Horizons | 0° | 20°S | 40°S | Zeniths | +0° | +20°S | +40°S | Ecliptic |

DEEP-SKY OBJECTS

Galaxy   Globular cluster   Open cluster   Diffuse nebula   Planetary nebula

STAR MAGNITUDES

-1   0   1   2   3   4   5   Variable star

WEST

SOUTHWEST

SOUTH

SOUTHEAST

EAST

SEXTANS
CRATER
CORVUS
HYDRA
ANTLIA
PYXIS
PUPPIS
VELA
CARINA
PICTOR
DORADO
RETICULUM
MENSA
LMC
VOLANS
CHAMAELEON
MUSCA
CRUX
Acrux
Becrux
Gacrux
CENTAURUS
NGC 5139
M83
α Centauri
β Centauri
CIRCINUS
APUS
OCTANS
HYDRUS
SMC
NGC 104
HOROLOGIUM
Achernar
ERIDANUS
PHOENIX
TUCANA
INDUS
PAVO
TRIANGULUM AUSTRALE
ARA
NORMA
LUPUS
LIBRA
M4
M80
Antares
SCORPIUS
M19
M62
M6
Shaula
M7
M9
M23
M21
M8
M18
M28
M24
M17
M22
M25
M54
M69
CORONA AUSTRALIS
SAGITTARIUS
M55
TELESCOPIUM
MICROSCOPIUM
CAPRICORNUS
M30
PISCIS AUSTRINUS
Fomalhaut
SCULPTOR

JULY

The northern night skies will continue to entice stargazers through the summer. High in the sky is Hercules, home to the magnificent M13. In the southern latitudes, the spectacular constellations Scorpius and Sagittarius are on show.

NORTHERN LATITUDES

THE STARS

The constellation Hercules is high in the sky this month and it is a good time to observe its celestial treasures, such as the globular cluster M13. Below M13 lies the winding constellation Draco, the Dragon. Towards the east

you can find the Summer Triangle asterism, while the constellation Ophiuchus sits in the south. Look for Boötes in the west, with the bright star Arcturus at its base. Below is Virgo and its brightest star Spica.

This is also an ideal time to observe the rich region of the sky covered by Scorpius and Sagittarius.

SIGHTS OF INTEREST

The globular cluster M13 in Hercules is a must-see object this month, as is another globular cluster, M5, which can be found in the nearby constellation Serpens. Ophiuchus also has some

interesting globular clusters, such as M10 and M12. These clusters are visible with binoculars, while a telescope will resolve many of their individual stars. If you have a pair of binoculars then look for the open clusters IC 4665 and NGC 6633, also in Ophiuchus.

LYRA

Size ranking	Brightest star	Genitive	Abbreviation	Highest in sky at 10 pm
52	Alpha (α) Lyrae, or Vega, 0.0	Lyrae	Lyr	July–August

You can easily find the relatively small constellation Lyra, the Lyre, by locating its brightest star Alpha (a) Lyrae, or Vega. Vega is one of the three stars of the famous Summer Triangle asterism. The planetary nebula M57, or the Ring Nebula, is also in Lyra and is a much-loved target for amateurs. Larger aperture telescopes will show the nebula as a small smoky, grey ring.

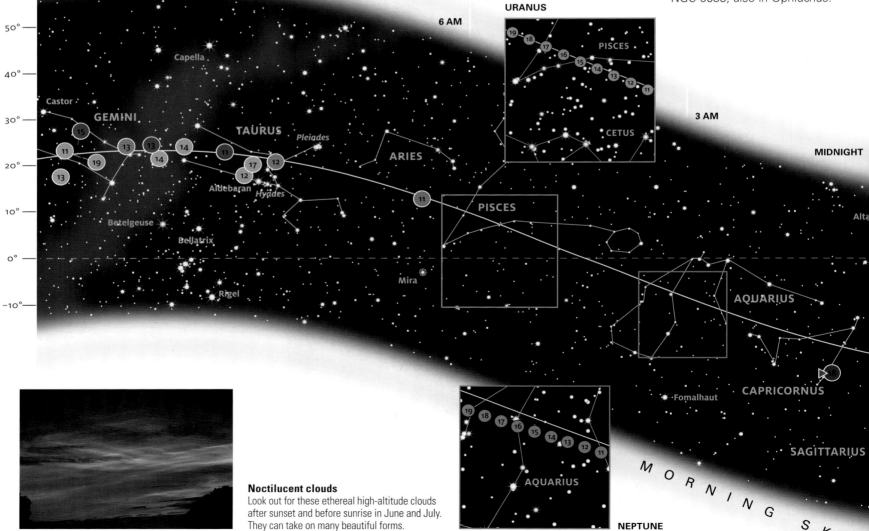

Noctilucent clouds
Look out for these ethereal high-altitude clouds after sunset and before sunrise in June and July. They can take on many beautiful forms.

SOUTHERN LATITUDES

THE STARS

Sitting high in the southern skies, Scorpius is easy to spot this month. Nearby are the constellations Sagittarius and the slightly less prominent Libra. Sagittarius is famous for the "Teapot" asterism formed by some of its brightest stars, and at this time it sits very high in the sky. When you look towards Sagittarius and Scorpius, you are peering towards the centre of the Milky Way galaxy. This whole region is full of rich and beautiful star fields, which are a joy to explore using binoculars.

A little lower in the sky are the bright stars Alpha (α) and Beta (β) Centauri, also known as Rigil Kentaurus and Hadar respectively. Very close to them you will spot the smallest constellation in the night sky – Crux, or the Southern Cross.

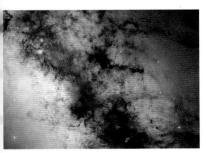

SIGHTS OF INTEREST

At this time of the year, the constellation Sagittarius offers some exceptional deep-sky objects. The striking globular cluster M22 is visible to the naked eye if you have good observing conditions. The Lagoon Nebula, or M8, lying above the spout of the "Teapot", is a glowing cloud of gas and a stunning sight through binoculars. It appears as a misty patch with the star cluster NGC 6530 nestled within it.

Other famous deep-sky objects in Sagittarius are visible through a telescope, including the Trifid Nebula, or M20. However, you can see one particularly bright patch of the Milky Way – M24 – with just the naked eye. Beside Sagittarius, Scorpius contains the bright open clusters M6 and M7, which remain high in the sky this month. To the north in the constellation Serpens Cauda, the Tail of the Serpent, lies the open cluster M16 in the much fainter Eagle Nebula.

The galactic hub
If you look towards the constellations Sagittarius and Scorpius on a clear night in the southern hemisphere, then you are looking in the direction of the very heart of our galaxy.

SAGITTARIUS

Size ranking	Brightest star	Genitive	Abbreviation	Highest in sky at 10 pm
15	Epsilon (ε) Sagittarii, 1.8	Sagittarii	Sgr	July–August

The constellation Sagittarius, the Archer, is nestled within a mesmerizingly detailed part of the Milky Way. You can find it by first locating the "Teapot" asterism, which forms the constellation's hub, close to a notably bright swathe of the Milky Way. A scan of Sagittarius with binoculars or a small telescope will reveal many rich star clusters and bright nebulae, such as the beautiful Lagoon Nebula.

METEOR SHOWER

When observing in late July, look out for the Delta Aquarid meteor shower, which peaks on 29 July. If observing from a dark sky site you should be able to spot around 20 meteors every hour.

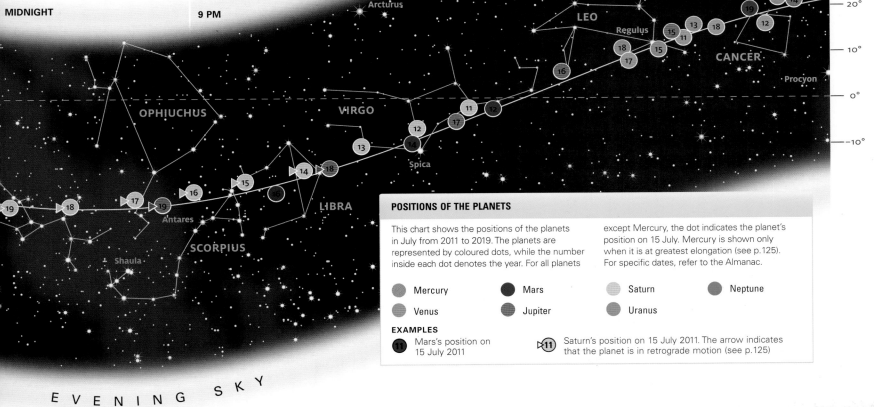

POSITIONS OF THE PLANETS

This chart shows the positions of the planets in July from 2011 to 2019. The planets are represented by coloured dots, while the number inside each dot denotes the year. For all planets except Mercury, the dot indicates the planet's position on 15 July. Mercury is shown only when it is at greatest elongation (see p.125). For specific dates, refer to the Almanac.

- Mercury
- Venus
- Mars
- Jupiter
- Saturn
- Uranus
- Neptune

EXAMPLES

⑪ Mars's position on 15 July 2011

▷⑪ Saturn's position on 15 July 2011. The arrow indicates that the planet is in retrograde motion (see p.125)

JULY
NORTHERN LATITUDES

OBSERVATION TIMES		
Date	Standard time	Daylight-saving time
15 June	Midnight	1 am
1 July	11 pm	Midnight
15 July	10 pm	11 pm
1 August	9 pm	10 pm
15 August	8 pm	9 pm

LOOKING NORTH

Ursa Major, the Great Bear, is home to several galaxies that can be seen with amateur equipment. M81, or Bode's Galaxy, appears as a fuzzy, grey blob through binoculars or a small telescope. The double star Alcor and Mizar is also worth a look while enjoying the sights of Ursa Major. In the east Cygnus, the Swan, looks magnificent at this time of the year. The more adventurous deep-sky observers should try to hunt down NGC 7000, the North America Nebula, just southeast of the star Deneb.

Alcor and Mizar
The double star system of Alcor and Mizar is visible to the naked eye. You can find it in the handle of the famous Plough asterism in the constellation Ursa Major.

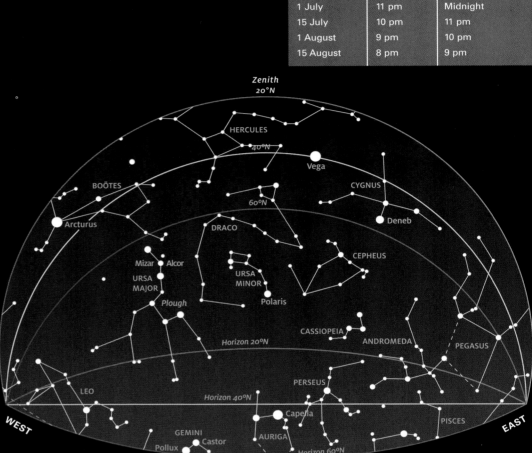

LOOKING SOUTH

One prominent marker of the night sky in the northern summer is the asterism known as the Summer Triangle. The corners of this large isosceles triangle are formed by the three bright stars Altair, Vega, and Deneb.

Another interesting sight is the multiple star system, the "Double-Double" (see p.86) or Epsilon (ε) Lyrae, in Lyra. Viewing it with binoculars shows a pair of stars, but, closer inspection with a telescope reveals that each of these stars is a pair of stars itself.

The Summer Triangle
This asterism is a useful navigational aid when finding your way around the summer night sky. Look for dark lanes in the Milky Way that cut across the triangle through Cygnus and beyond.

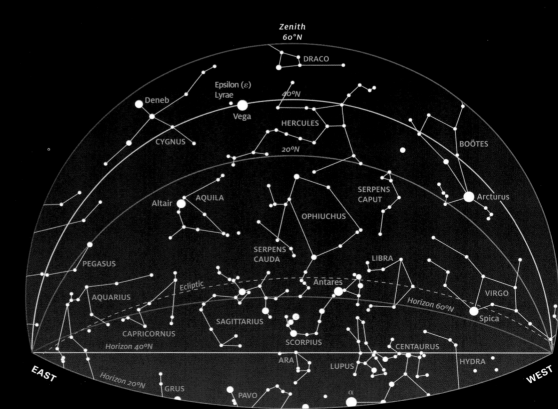

STAR MAGNITUDES

● -1   ● 0   ● 1   ● 2   • 3 and above

JULY
SOUTHERN LATITUDES

LOOKING **NORTH**

The wonderful globular cluster M5 is high in the sky at this time. It is roughly 25,000 light-years away from Earth, towards the constellation Serpens Caput. A small telescope brings many of its outer stars into focus. A short hop east over Ophiuchus into Serpens Cauda and you will find the open cluster M16 surrounded by the much fainter Eagle Nebula. This cluster can be observed with a pair of binoculars. Also look out for Alpha (α) Librae, a double star in the constellation Libra, the Scales.

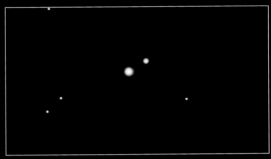

Alpha (α) Librae
The second brightest star in Libra is called Zubenelgenubi (Arabic for "the Southern claw"). It is a binary star system, and a pair of binoculars easily shows its two stars.

LOOKING **SOUTH**

The "Teapot" asterism, made of eight stars, lies among the stars of the Milky Way in the constellation Sagittarius. Its stubby spout is marked by the stars Gamma (γ), Epsilon (ε), and Delta (δ) Sagittarii, while Phi (φ), Sigma (σ), Zeta (ζ), and Tau (τ) make its handle. Scattered around the Teapot are some interesting binocular and small-telescope targets including the bright Lagoon Nebula M8, the magnitude 4.6 open cluster M25, and the globular cluster M22.

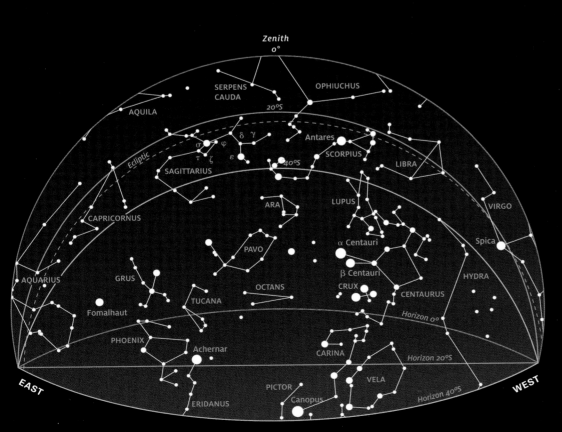

M22 in Sagittarius
A small telescope shows many of M22's brightest stars. It is the third-brightest globular cluster in the sky and can be seen with just the naked eye in particularly dark and clear skies.

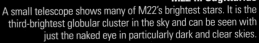

JULY | NORTHERN LATITUDES

LOOKING NORTH

STAR MAGNITUDES

* -1
* 0
* 1
* 2
* 3
* 4
* 5
* Variable star

DEEP-SKY OBJECTS

* Galaxy
* Globular cluster
* Open cluster
* Diffuse nebula
* Planetary nebula

POINTS OF REFERENCE

| Horizons | 60°N | 40°N | 20°N | Zeniths | 60°N | 40°N | 20°N | Ecliptic |

OBSERVATION TIMES

Date	Standard time	Daylight-saving time
15 June	Midnight	1 am
1 July	11 pm	Midnight
15 July	10 pm	11 pm
1 August	9 pm	10 pm
15 August	8 pm	9 pm

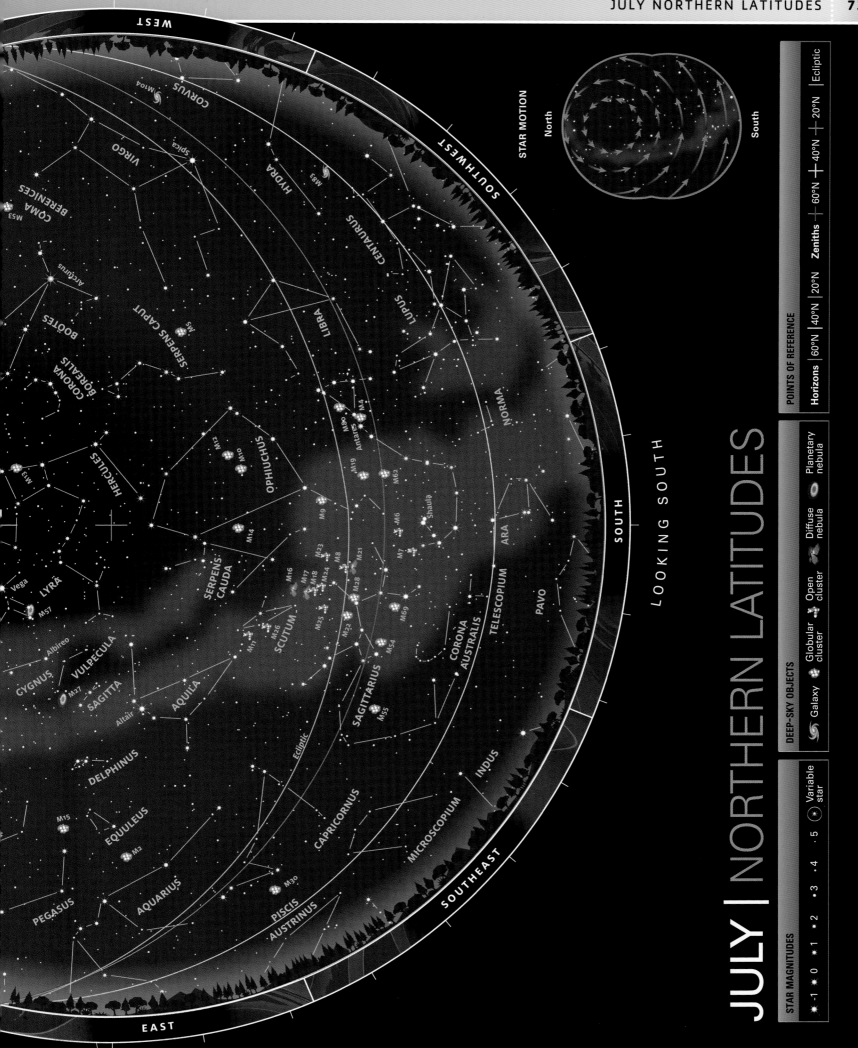

JULY | NORTHERN LATITUDES

LOOKING SOUTH

STAR MOTION

North

South

POINTS OF REFERENCE

Horizons | 60°N | 40°N | 20°N
Zeniths | + 60°N | + 40°N | + 20°N | Ecliptic

DEEP-SKY OBJECTS

Galaxy | Globular cluster | Open cluster | Diffuse nebula | Planetary nebula

STAR MAGNITUDES

-1 | 0 | 1 | 2 | 3 | 4 | 5 | Variable star

Constellation and object labels

WEST
EAST
SOUTH
SOUTHWEST
SOUTHEAST

CORVUS
M104
VIRGO
Spica
COMA BERENICES
M53
BOÖTES
Arcturus
CORONA BOREALIS
HERCULES
M13
SERPENS CAPUT
M5
HYDRA
M83
CENTAURUS
LIBRA
LUPUS
NORMA
OPHIUCHUS
M12
M10
M19
M90
Antares
M4
M62
Shaula
ARA
M14
M9
SERPENS CAUDA
M16
M17
M23
M18
M24
M8
M28
M21
M7
M6
M25
M26
SCUTUM
M11
M22
M69
M54
M55
SAGITTARIUS
CORONA AUSTRALIS
TELESCOPIUM
PAVO
INDUS
Vega
LYRA
M57
Albireo
CYGNUS
VULPECULA
M27
SAGITTA
AQUILA
Altair
DELPHINUS
CAPRICORNUS
MICROSCOPIUM
EQUULEUS
M15
M2
PEGASUS
AQUARIUS
M30
PISCIS AUSTRINUS
Ecliptic

JULY | SOUTHERN LATITUDES

STAR MAGNITUDES

* -1
* 0
* 1
* 2
* 3
* 4
* 5
* Variable star

DEEP-SKY OBJECTS

Galaxy
Globular cluster
Open cluster
Diffuse nebula
Planetary nebula

POINTS OF REFERENCE

| Horizons | 0° | 20°S | 40°S | |
| Zeniths | 0° | 20°S | 40°S | Ecliptic |

LOOKING NORTH

OBSERVATION TIMES		
Date	Standard time	Daylight-saving time
15 June	Midnight	1 am
1 July	11 pm	Midnight
15 July	10 pm	11 pm
1 August	9 pm	10 pm
15 August	8 pm	9 pm

JULY | SOUTHERN LATITUDES

LOOKING SOUTH

STAR MOTION

North   South

STAR MAGNITUDES

-1   0   1   2   .3   .4   .5   ⊙ Variable star

DEEP-SKY OBJECTS

Galaxy   Globular cluster   Open cluster   Diffuse nebula   Planetary nebula

POINTS OF REFERENCE

Horizons   0°   20°S   40°S
Zeniths   0°   20°S   40°S
Ecliptic

WEST

SOUTHWEST

SOUTH

SOUTHEAST

EAST

CORVUS
VIRGO
Spica
CRATER
HYDRA
ANTLIA
VELA
CENTAURUS
M104
M83
NGC 5139
α Centauri
β Centauri
Becrux
Gacrux
Acrux
CRUX
MUSCA
CIRCINUS
CARINA
VOLANS
CHAMAELEON
MENSA
PICTOR
Canopus
LMC
DORADO
RETICULUM
HYDRUS
HOROLOGIUM
ERIDANUS
Achernar
SMC
NGC 104
TUCANA
PHOENIX
SCULPTOR
GRUS
INDUS
PAVO
OCTANS
APUS
TRIANGULUM AUSTRALE
ARA
NORMA
LUPUS
LIBRA
SCORPIUS
Antares
M4
M80
M19
M62
M6
M7
Shaula
TELESCOPIUM
CORONA AUSTRALIS
SAGITTARIUS
M54
M69
M22
M55
M28
M8
M21
M20
MICROSCOPIUM
CAPRICORNUS
M30
PISCIS AUSTRINUS
Fomalhaut
AQUARIUS
Ecliptic

AUGUST

On a warm August evening the most prominent feature from northern latitudes is the large Summer Triangle asterism. From the southern hemisphere, the magnificent centre of the Milky Way is still sitting high in the sky.

NORTHERN LATITUDES

THE STARS

Directly overhead is the bright star Vega in Lyra, as well as Deneb, which marks the tail of the constellation Cygnus. The shape of Cygnus means it is often called the Northern Cross. In the south, the rich regions around Scutum, Scorpius, and Sagittarius are sinking away.

SIGHTS OF INTEREST

When observing Cygnus, look out for the Cygnus Rift. This dark lane of dust sits in front of the background stars and appears to split the Milky Way into two. Also look out for the Wild Duck Cluster, M11, in Scutum. It is clearly visible through binoculars.

CYGNUS

Size ranking	Brightest star	Genitive	Abbreviation	Highest in sky at 10 pm
16	Alpha (α) Cygni, or Deneb 1.3	Cygni	Cyg	August–September

The constellation Cygnus, the Swan, is easy to recognize due to its large "cross" shape. Its brightest star Deneb, has magnitude 1.3, and marks the swan's tail. The swan's head is marked by the gorgeous double star Albireo, Beta (β) Cygni. A small telescope reveals the two stars – one gold, the other tinted blue. Binoculars are perfect for exploring Cygnus's beautiful star fields and clusters.

METEOR SHOWER

One of the finest meteor showers of the year, the Perseids peak around 12 August. This is a great opportunity to lie back, take in the night sky, and hopefully see some meteors too – you should be able to spot one every minute or so. The meteors appear to come from the northern parts of Perseus. They are typically quite bright and are best seen after midnight.

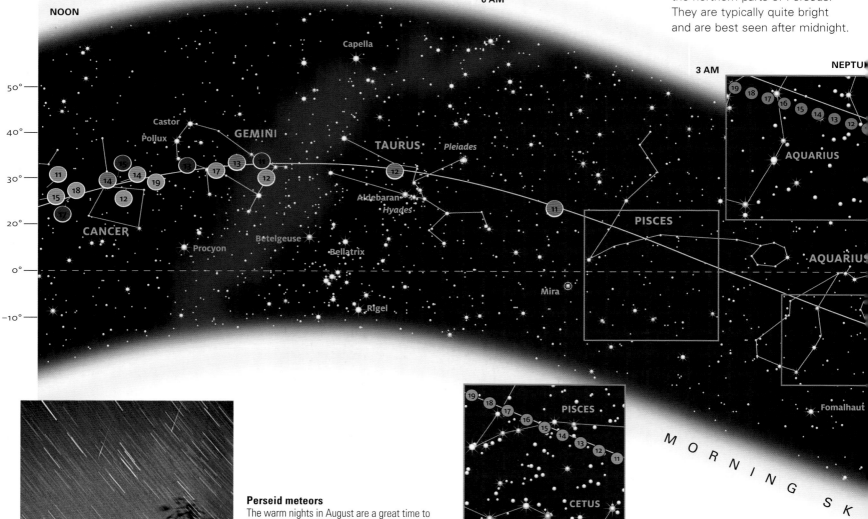

Perseid meteors
The warm nights in August are a great time to sit back and watch the Perseid meteor shower unfold in the night sky above you.

SOUTHERN LATITUDES

THE STARS

When observing from southern latitudes you can find Sagittarius, the Archer, lying almost overhead; to its southwest lies Scorpius, the Scorpion. Low on the southwest the bright stars Alpha (α) and Beta (β) Centauri, also known as Rigil Kentaurus and Hadar, are sinking below the horizon, taking Centaurus with them.

Low on the horizon between the stars of Centaurus and Scorpius is the constellation Lupus, the Wolf. In the east the bright star Fomalhaut lies in the constellation Piscis Austrinus. Between Fomalhaut and the stars of Scorpius are the stars of the constellations Grus, Tucana, Pavo, and Ara. With clear skies you should have little trouble seeing the Small Magellanic Cloud sitting to the west of the star Achernar in Eridanus.

SIGHTS OF INTEREST

With the rich regions of Sagittarius and Scutum visible high in the sky, you can take your pick from a superb selection of binocular and telescope targets this month. To experience a real space-walking feel, pick up a pair of binoculars and "wander" along the Milky Way, from Scutum to Centaurus.

If you have a telescope, the Lagoon Nebula makes a wonderful target in Sagittarius, and with a wide enough field of view you can also glimpse the Trifid Nebula, M20. Two interesting open clusters M6 and M7 are nestled among rich star fields in Scorpius, and both are visible to the naked eye. The star field M24 in Sagittarius makes a great binocular target. Looking northwards, you will find the planetary nebula M57, or the Ring Nebula. It is an interesting target for a small telescope, as is the larger planetary nebula M27, or the Dumbbell Nebula, in the constellation Vulpecula, the Fox.

SCUTUM

Size ranking	Brightest star	Genitive	Abbreviation	Highest in sky at 10 pm
84	Alpha (α) Scuti, 3.8	Scuti	Sct	July–August

The constellation Scutum, the Shield, is relatively small – the fifth smallest of 88 constellations. It is located between the stars of Aquila and Sagittarius, quite close to the constellation Serpens Cauda, the Snake's Tail, in a wonderfully rich and interesting part of the Milky Way. The Polish astronomer Johannes Hevelius originally named it "Sobieski's Shield" in 1684, in honour of John Sobieski, the king of Poland at that time.

The Lagoon Nebula in Sagittarius
M8, or the Lagoon Nebula (bottom right), can be seen with just the naked eye and makes an excellent target for a small telescope. It appears nestled among the rich star fields of our galaxy, the Milky Way.

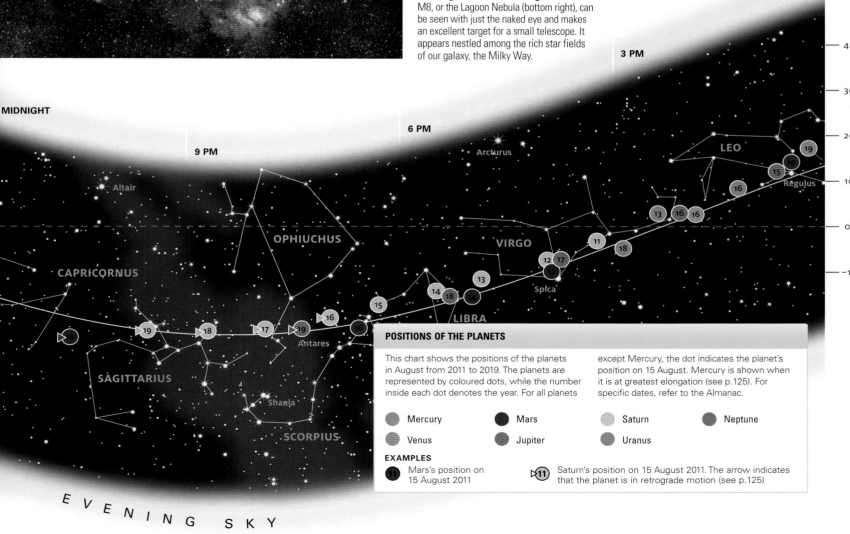

POSITIONS OF THE PLANETS

This chart shows the positions of the planets in August from 2011 to 2019. The planets are represented by coloured dots, while the number inside each dot denotes the year. For all planets except Mercury, the dot indicates the planet's position on 15 August. Mercury is shown when it is at greatest elongation (see p.125). For specific dates, refer to the Almanac.

- ○ Mercury
- ○ Venus
- ● Mars
- ● Jupiter
- ● Saturn
- ● Uranus
- ● Neptune

EXAMPLES

- ⑪ Mars's position on 15 August 2011
- ⑪ Saturn's position on 15 August 2011. The arrow indicates that the planet is in retrograde motion (see p.125)

EVENING SKY

AUGUST
NORTHERN LATITUDES

OBSERVATION TIMES		
Date	Standard time	Daylight-saving time
15 July	Midnight	1 am
1 August	11 pm	Midnight
15 August	10 pm	11 pm
1 September	9 pm	10 pm
15 September	8 pm	9 pm

LOOKING NORTH

Northern summer nights are the perfect time to admire the galaxy we live in. In August the Milky Way rises up from Auriga and Perseus in the northeast, stretching all the way across the sky into Scutum, Scorpius, and Sagittarius in the southwest. There are many objects nestled in and around the Milky Way that can be spotted with binoculars. Besides the Double Cluster (see p.22) be sure to look out for the globular clusters M13 and M92. A large telescope will reveal the galaxies M81 and M51.

The Milky Way
Binoculars are the ideal way to explore the star fields of the Milky Way. The glow of the Milky Way on a dark night is the collective light of billions of the galaxy's stars.

LOOKING SOUTH

There are two must-see objects if you are looking south in the northern hemisphere: M27, or the Dumbbell Nebula, east of Pegasus, and M57, or the Ring Nebula, south of Cygnus. Both these planetary nebulae are enormous shells of gas ejected by Sun-like stars as they die. The Ring Nebula is so named because it appears like a smoky grey ring when it is viewed through a telescope. The Dumbbell Nebula appears as a faint grey bow-tie shape through a large-aperture telescope.

The Dumbbell Nebula
Lying east of Pegasus, the Dumbbell Nebula can be seen as a fuzzy patch through a small telescope or a pair of binoculars. A large telescope reveals its intriguing shape more clearly.

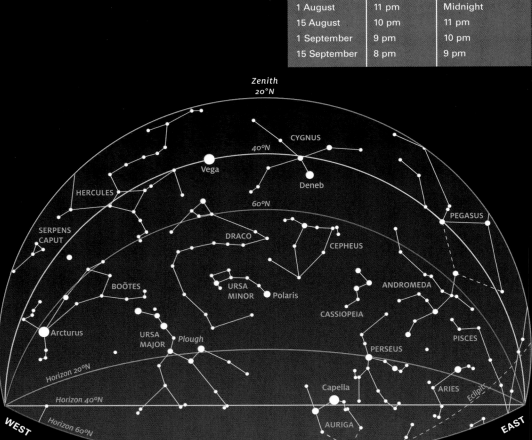

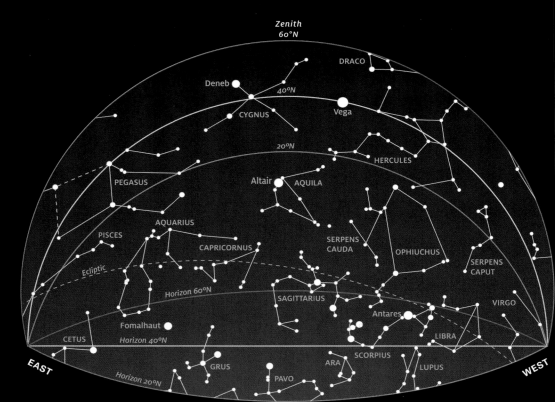

AUGUST
SOUTHERN LATITUDES

AUGUST

STAR MAGNITUDES
●-1 ●0 ●1 ●2 •3 and above

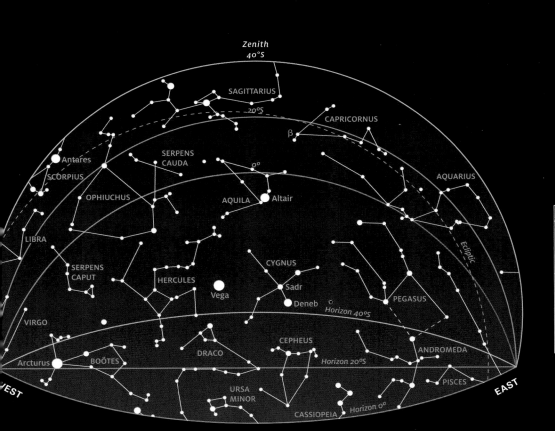

Zenith
40°S

LOOKING **NORTH**

The constellation Cygnus, the Swan, is home to two open clusters that make good small-telescope targets. M29 sits very close to the star Sadr, Gamma (γ) Cygni. The magnitude 4.6 cluster M39 can be seen sparkling against the stars of the Milky Way. It has about 30 stars. Higher in the sky you will find Capricornus, the Sea Goat, which is home to the globular cluster M30 and Beta (β) Capricorni – a double star of magnitude 3.1 that can be seen with binoculars.

M39 in Cygnus
The open cluster M39 covers an area of similar size to the full Moon and lies 825 light-years away. It is a nice target for binoculars or a small telescope on a clear night.

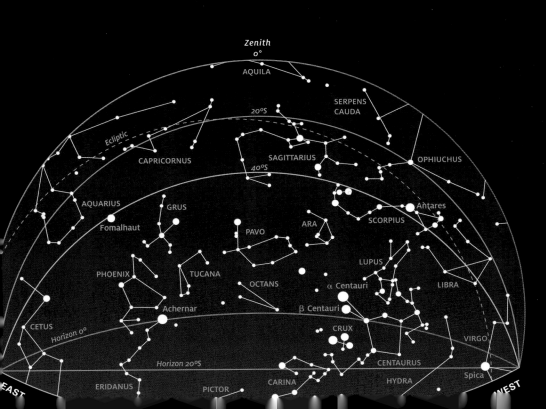

Zenith
0°

LOOKING **SOUTH**

Sagittarius, the Archer, is now sitting high in the south, providing a feast of objects to observe. M17, the Omega Nebula, is a good target for a small telescope. It is a glowing cloud of hydrogen gas, which resembles the Greek capital letter Omega (ω). The open cluster M23 and the Sagittarius Star Cloud M24 are also ideal binocular objects. A large telescope is needed to clearly see the magnitude 9 M20, also known as the Trifid Nebula.

The Trifid Nebula
The Trifid Nebula is an enormous cloud of gas 7,600 light-years from Earth. It lies in the constellation Sagittarius, and at its heart

AUGUST | NORTHERN LATITUDES

LOOKING NORTH

STAR MAGNITUDES

✦	-1
✦	0
✷	1
·	2
·	3
·	4
·	5
⊙	Variable star

DEEP-SKY OBJECTS

🌀	Galaxy
⊛	Globular cluster
✳	Open cluster
✺	Diffuse nebula
⬭	Planetary nebula

POINTS OF REFERENCE

Horizons	60°N	40°N	20°N
Zeniths	+60°N	+40°N	+20°N	Ecliptic

OBSERVATION TIMES

Date	Standard time	Daylight-saving time
15 July	Midnight	1 am
1 August	11 pm	Midnight
15 August	10 pm	11 pm
1 September	9 pm	10 pm
15 September	8 pm	9 pm

AUGUST | NORTHERN LATITUDES

LOOKING SOUTH

STAR MOTION

North

South

POINTS OF REFERENCE

Horizons | 60°N | 40°N | 20°N | Zeniths + 60°N + 40°N + 20°N | Ecliptic

DEEP-SKY OBJECTS

Galaxy | Globular cluster | Open cluster | Diffuse nebula | Planetary nebula

STAR MAGNITUDES

-1 · 0 · 1 · 2 · 3 · 4 · 5 | Variable star

AUGUST | SOUTHERN LATITUDES

LOOKING NORTH

STAR MAGNITUDES

- ★ -1
- ★ 0
- ✳ 1
- ✳ 2
- · 3
- · 4
- · 5
- ⊛ Variable star

DEEP-SKY OBJECTS

- 🌀 Galaxy
- ✴ Globular cluster
- ✴ Open cluster
- ❀ Diffuse nebula
- ⬭ Planetary nebula

POINTS OF REFERENCE

Horizons | 0° | 20°S | 40°S

Zeniths — 0° — 20°S — 40°S

Ecliptic

OBSERVATION TIMES

Date	Standard time	Daylight-saving time
15 July	Midnight	1 am
1 August	11 pm	Midnight
15 August	10 pm	11 pm
1 September	9 pm	10 pm
15 September	8 pm	9 pm

AUGUST | SOUTHERN LATITUDES

LOOKING SOUTH

WEST

SOUTHWEST

SOUTH

SOUTHEAST

EAST

WEST

STAR MOTION

North

South

STAR MAGNITUDES

-1 · 0 · 1 · 2 · 3 · 4 · 5 · Variable star

DEEP-SKY OBJECTS

Galaxy · Globular cluster · Open cluster · Diffuse nebula · Planetary nebula

POINTS OF REFERENCE

Horizons | 0° | 20°S | 40°S · Zeniths + 0° + 20°S + 40°S · Ecliptic

Constellations and objects

VIRGO
Spica
Ecliptic
LIBRA
HYDRA
M83
NGC 5139
CENTAURUS
LUPUS
CIRCINUS
NORMA
SCORPIUS
Shaula
Antares
M4
M80
M19
M62
M6
M7
M9
M21
M8
M20
M28
M22
M54
M69
SAGITTARIUS
CORONA AUSTRALIS
M55
M30
ARA
TELESCOPIUM
PAVO
APUS
OCTANS
TRIANGULUM AUSTRALE
α Centauri
β Centauri
Becrux
Gacrux
CRUX
Acrux
MUSCA
CHAMAELEON
VELA
CARINA
Canopus
PICTOR
VOLANS
MENSA
LMC
DORADO
RETICULUM
HYDRUS
HOROLOGIUM
SMC
NGC 104
TUCANA
GRUS
Achernar
PHOENIX
ERIDANUS
FORNAX
SCULPTOR
CETUS
INDUS
MICROSCOPIUM
CAPRICORNUS
AQUARIUS
PISCIS AUSTRINUS
Fomalhaut
M30

SEPTEMBER

The nights are darker now in the northern hemisphere, making it a good time to admire the constellations along the Milky Way. In the southern hemisphere, the region around the Milky Way's centre moves to the west.

NORTHERN LATITUDES

THE STARS

Looking high up in the sky, you will spot the constellation Cepheus, representing King Cepheus. Delta (δ) Cephei, a variable star in Cepheus, is a popular target for amateur astronomers. Its brightness varies between magnitude 3.5 and 4.4 every 5 days and 9 hours.

Towards the west, the stars of the Summer Triangle are still on show, while the constellations Cassiopeia and Andromeda are visible in the east. The roughly triangular constellation Capricornus, the Sea Goat, lies in the south.

SIGHTS OF INTEREST

If you are up for a real challenge this month, try hunting down the North America Nebula, NGC 7000. It is hard to detect from light polluted skies, but with binoculars it can be seen from dark skies, sitting near the star Deneb in Cygnus. If you want to marvel at one of the jewels of the night sky, look out for the beautiful globular cluster M15 using binoculars. It can be seen near the star Enif, or Epsilon (ε) Pegasi. The open star cluster M39 in Cygnus is another deep-sky object worth observing through binoculars or a small telescope.

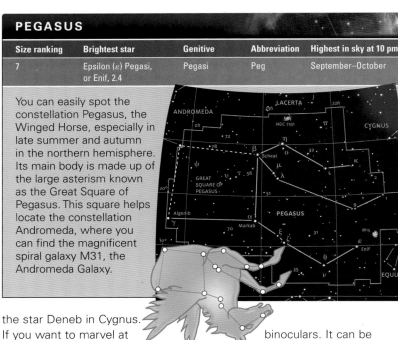

PEGASUS				
Size ranking	Brightest star	Genitive	Abbreviation	Highest in sky at 10 pm
7	Epsilon (ε) Pegasi, or Enif, 2.4	Pegasi	Peg	September–October

You can easily spot the constellation Pegasus, the Winged Horse, especially in late summer and autumn in the northern hemisphere. Its main body is made up of the large asterism known as the Great Square of Pegasus. This square helps locate the constellation Andromeda, where you can find the magnificent spiral galaxy M31, the Andromeda Galaxy.

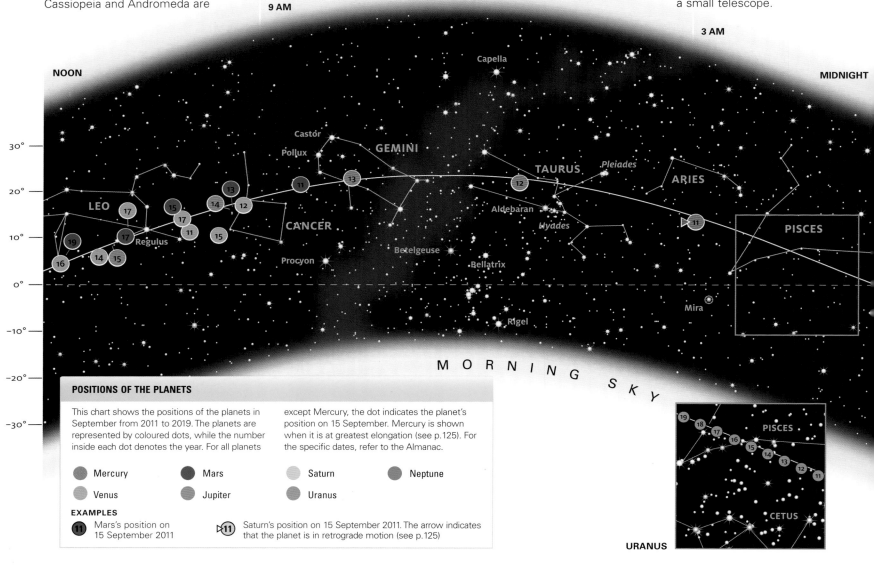

POSITIONS OF THE PLANETS

This chart shows the positions of the planets in September from 2011 to 2019. The planets are represented by coloured dots, while the number inside each dot denotes the year. For all planets except Mercury, the dot indicates the planet's position on 15 September. Mercury is shown when it is at greatest elongation (see p.125). For the specific dates, refer to the Almanac.

- Mercury
- Venus
- Mars
- Jupiter
- Saturn
- Uranus
- Neptune

EXAMPLES

(11) Mars's position on 15 September 2011

(▷11) Saturn's position on 15 September 2011. The arrow indicates that the planet is in retrograde motion (see p.125)

URANUS

SOUTHERN LATITUDES

EVENING SKY

NEPTUNE

AQUARIUS

SCORPIUS

SAGITTARIUS

LIBRA

λ Shaula

Antares

OPHIUCHUS

CAPRICORNUS

VIRGO

AQUARIUS

Altair

Arcturus

NOON · 3 PM · 6 PM · 9 PM

THE STARS

If you are observing from the southern hemisphere this month, be sure to enjoy the rich regions of Scorpius and Ophiuchus before they sink below the western horizon. Just above them lies a rich region around the heart of the Milky Way galaxy, brimming with star clusters and bright nebulae.

In contrast, the eastern half of the sky is relatively empty, though you can still find several constellations, including Pisces, the Fishes; Cetus, the Whale; and Eridanus, the River.

SIGHTS OF INTEREST

This month be sure to enjoy the sights of Scorpius, Sagittarius, and Scutum before they disappear below the horizon. The most spectacular objects to look out for in these constellations include M8, the Lagoon Nebula; the open clusters M6 and M7; and the globular cluster M22. The constellation Aquarius, the Water Carrier, sits almost overhead. You can see several deep-sky objects there, including the interesting planetary nebula NGC 7293, also known as the Helix Nebula. You will need dark skies and a relatively large telescope to view this nebula. There are two interesting binocular targets in the sky at the moment – the globular cluster M2 in Aquarius, near the star Beta Aquarii, and another globular cluster M15, in Pegasus.

The Small Magellanic Cloud
Lying in the constellation Tucana in the southern hemisphere, the Small Magellanic Cloud can be found close to the beautiful globular cluster 47 Tucanae, also known as NGC 104.

PISCIS AUSTRINUS

Brightest star	Size ranking	Genitive	Abbreviation	Highest in sky at 10 pm
Alpha (α) Piscis Austrini, or Fomalhaut, 1.2	60	Piscis Austrini	PsA	September–October

Also known as the Southern Fish, Piscis Austrinus is one of the smaller constellations in the night sky and lacks any prominent deep-sky objects. You can find it nestled between the constellations Grus, Aquarius, Capricornus, and Sculptor. Its brightest star is the blue-white coloured Fomalhaut, which sits at the mouth of the fish. Fomalhaut lies at a distance of 25 light-years from Earth.

SEPTEMBER
NORTHERN LATITUDES

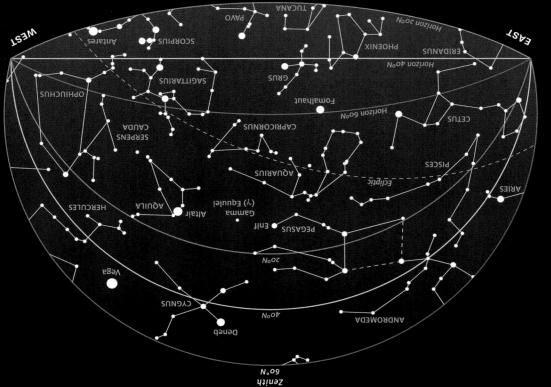

Zenith
20°N

OBSERVATION TIMES		
Date	Standard time	Daylight-saving time
15 August	Midnight	1 am
1 September	11 pm	Midnight
15 September	10 pm	11 pm
1 October	9 pm	10 pm
15 October	8 pm	9 pm

LOOKING NORTH

In the northern skies, trace the constellation Draco, the Dragon, weaving between the constellations Ursa Minor, Cepheus, and Hercules. At the tip of the dragon's tongue lies the double star 16 and 17 Draconis, which can be spotted with just a pair of binoculars. Nu (ν) Draconis in the dragon's head is also an interesting double star in binoculars. Sitting a little way above the bright star Vega, the multiple star system Epsilon (ε) Lyrae, is an excellent target for a telescope.

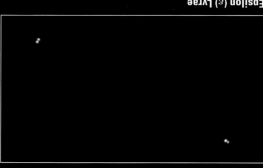

Epsilon (ε) Lyrae
Also known as the "Double-Double", with binoculars this quadruple star appears as a pair, while a small telescope shows the two stars' twin components.

LOOKING SOUTH

In the south the impressive Andromeda Galaxy, M31, sits in the heart of Andromeda. It is easily visible to the naked eye from a dark sky site. Binoculars reveal it as a fuzzy ellipse, while a small telescope shows it more clearly.

The Coathanger asterism, or Brocchi's Cluster, lies between Cygnus and Aquila in the east, and is easily visible through binoculars. The double star Gamma (γ) Equulei, west of the constellation Aquila, sits near the star Enif in Pegasus and is also a good binocular target.

The Coathanger
Ten stars make up the famous shape of the Coathanger, south of Cygnus. The stars of this open cluster are not near each other in space; the shape is a chance alignment.

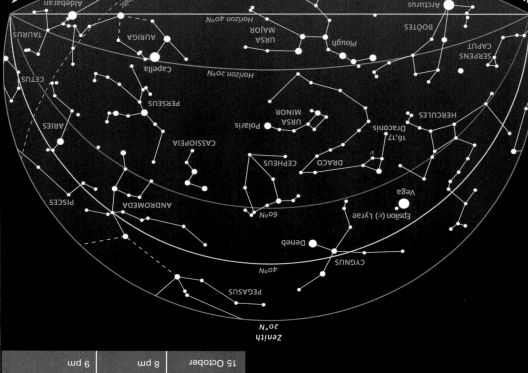

Zenith
60°N

SEPTEMBER
SOUTHERN LATITUDES

LOOKING NORTH

The globular cluster M15 sits southwest of the star Enif, Epsilon (ε) Pegasi, in Pegasus. This dense cluster can be picked out with binoculars, and a small telescope shows it clearly. M15 is thought to be 13.2 billion years old. Several deep-sky objects lie in the northeast, in the constellation Aquarius. The globular cluster M2 appears as a fuzzy star through binoculars, and the planetary nebula NGC 7293, the Helix Nebula, appears as a faint fuzzy disc through a small telescope.

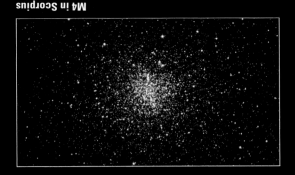

M15 in Pegasus
The globular cluster M15 is roughly 175 light-years in diameter and 30,000 light-years from Earth. A 150mm (6in) telescope reveals many of the cluster's sparkling stars.

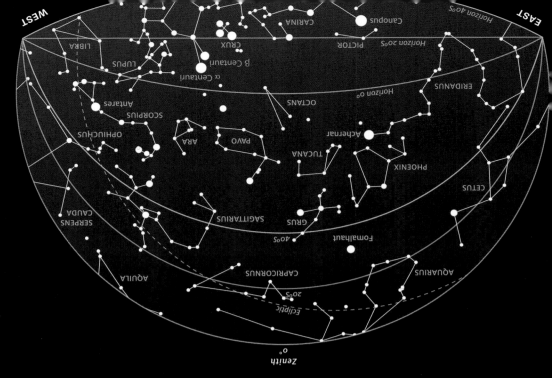

LOOKING SOUTH

The globular cluster 47 Tucanae is a must-see target in the September southern skies, lying to the south of the constellation Tucana. It is visible to the naked eye as a hazy star, while a small telescope shows its bright centre and many of its glittering stars. This cluster is 15,000 light-years away. Other visible targets include the globular clusters M22, NGC 6397, and M4, in Sagittarius, Ara, and Scorpius respectively. The open clusters M6 and M7 in Scorpius are also visible.

M4 in Scorpius
The globular cluster M4 is a beautiful sight in the constellation Scorpius. Sitting close to the star Antares, Alpha (α) Scorpii, it is a lovely target for binoculars or a small telescope.

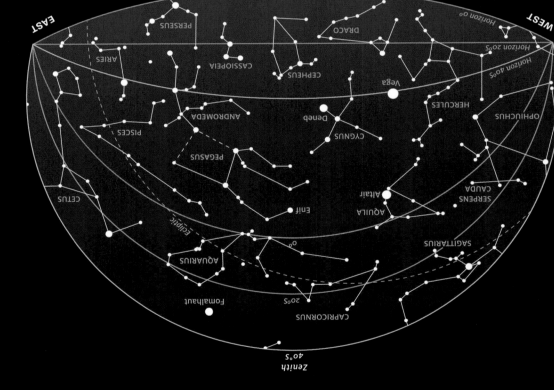

SEPTEMBER | NORTHERN LATITUDES

LOOKING NORTH

OBSERVATION TIMES

Date	Standard time	Daylight-saving time
15 August	Midnight	1 am
1 September	11 pm	Midnight
15 September	10 pm	11 pm
1 October	9 pm	10 pm
15 October	8 pm	9 pm

NORTHEAST

NORTH

NORTHWEST

WEST

TAURUS
Aldebaran
Hyades
M45 (Pleiades)
ARIES
TRIANGULUM
M33
PERSEUS
M34
NGC 869
NGC 884
ANDROMEDA
M31
CASSIOPEIA
M52
LACERTA
M39
Deneb
CYGNUS
CEPHEUS
Polaris
URSA MINOR
DRACO
LYRA
Vega
M57
M92
HERCULES
M13
CORONA BOREALIS
BOÖTES
Arcturus
SERPENS CAPUT
COMA BERENICES
M3
CANES VENATICI
M101
Mizar
M51
Plough
URSA MAJOR
LEO MINOR
M81
LYNX
CAMELOPARDALIS
AURIGA
Capella
M38
M36
M37
GEMINI
Castor
M1

STAR MAGNITUDES

- ✦ -1
- ✦ 0
- ✦ 1
- ∙ 2
- · 3
- · 4
- · 5
- ◉ Variable star

DEEP-SKY OBJECTS

- 🌀 Galaxy
- ✸ Globular cluster
- ✳ Open cluster
- ✦ Diffuse nebula
- ⬭ Planetary nebula

POINTS OF REFERENCE

Horizons	60°N	40°N	20°N
Zeniths	+ 60°N	+ 40°N	+ 20°N
			Ecliptic

LOOKING SOUTH

STAR MOTION

North

South

SEPTEMBER | SOUTHERN LATITUDES

LOOKING NORTH

OBSERVATION TIMES

Date	Standard time	Daylight-saving time
15 August	Midnight	1 am
1 September	11 pm	Midnight
15 September	10 pm	11 pm
1 October	9 pm	10 pm
15 October	8 pm	9 pm

POINTS OF REFERENCE

Horizons	0°	20°S	40°S	Zeniths	0°	20°S	40°S	Ecliptic

STAR MAGNITUDES

-1 · 0 · 1 · 2 · 3 · 4 · 5 ⊙ Variable star

DEEP-SKY OBJECTS

Galaxy · Globular cluster · Open cluster · Diffuse nebula · Planetary nebula

SEPTEMBER | SOUTHERN LATITUDES

STAR MAGNITUDES

- -1
- 0
- 1
- 2
- 3
- 4
- 5
- Variable star

DEEP-SKY OBJECTS

- Galaxy
- Globular cluster
- Open cluster
- Diffuse nebula
- Planetary nebula

POINTS OF REFERENCE

Horizons | 0° | 20°S | 40°S

Zeniths + 0° + 20°S + 40°S | Ecliptic

LOOKING SOUTH

STAR MOTION

South

North

OCTOBER

In the northern hemisphere, the main attractions this month are Pegasus and Andromeda. In the southern hemisphere, these constellations sit in the north, while the Small Magellanic Cloud lies in the south.

NORTHERN LATITUDES

THE STARS

Standing under the cold, crisp October night sky, you can see the Great Square of Pegasus high up in the sky. You can easily spot the constellation Andromeda sitting next to it, while a loop of stars known as the Circlet is visible directly beneath the asterism. The constellations Aquarius, Pisces, and Cetus can be located nearby. Turn around and look towards the north to find the constellations Cepheus, Cassiopeia, and Perseus on show. Cygnus, the Swan, and Lyra, the Lyre, are visible towards the west. Although a small constellation, Lyra can be found quite easily by locating its brightest star, the magnitude 0.0 Vega.

SIGHTS OF INTEREST

If you have a small telescope or a pair of binoculars the Andromeda Galaxy, or M31, in Andromeda is a wonderful target to look at. M31 is visible to the naked eye from dark skies. Binoculars show the open cluster M52 in Cassiopeia.

METEOR SHOWER

The Orionid meteor shower reaches its peak around 21 October. Under good conditions during the peak you can expect to see around 25 meteors every hour, shooting from the direction of the patch of sky between Orion's head and Gemini's feet. The best time to look is after midnight, when this region of sky has risen higher above the horizon.

(Star chart labels: MORNING SKY — Mira, Rigel, Bellatrix, Betelgeuse, Procyon, ARIES, Hyades, Aldebaran, TAURUS, Pleiades, GEMINI, Castor, Pollux, CANCER, LEO, Regulus, VIRGO, Capella, NOON, 9 AM, 6 AM, 3 AM, MIDNIGHT)

PERSEUS

Brightest star	Genitive	Abbreviation	Size ranking	Highest in sky at 10 pm
Alpha (α) Persei, or Mirphak, 1.8	Persei	Per	24	November–December

Perseus, the Hero, sits among the beautiful star fields of the Milky Way Galaxy between Andromeda and Auriga. It is an interesting constellation that contains some wonderful deep-sky objects to observe with a small telescope or a pair of binoculars, including the open cluster M34 and the Double Cluster, NGC 884 and NGC 869.

POSITIONS OF THE PLANETS

This chart shows the positions of the planets in October from 2011 to 2019. The planets are represented by coloured dots, while the number inside each dot denotes the year. For all planets except Mercury, the dot indicates the planet's position on 15 October. Mercury is shown when it is at greatest elongation (see p.125). For specific dates, refer to the Almanac.

- Mercury
- Venus
- Mars
- Jupiter
- Uranus
- Saturn
- Neptune

EXAMPLES

- Mars's position on 15 October 2011
- Saturn's position on 15 October 2011. The arrow indicates that the planet is in retrograde motion (see p.125)

SOUTHERN LATITUDES

THE STARS

After the wonderfully rich views of the southern winter, October night skies look rather empty. However, this does not mean there is nothing to see. Looking south, you can locate the constellations Phoenix, Grus, Tucana, Pavo, and the long and winding Eridanus. The constellation Sagittarius, the Archer, lies low in the west.

There are several bright stars to help you navigate the sky. In the south, look out for the magnitude 1.2 star Fomalhaut, almost directly above you in the constellation Piscis Austrinus. A little lower down in the south at one end of Eridanus shines Achernar, or Alpha (α) Eridani, while the bright star Altair, in Aquila, twinkles away in the west. Aquarius is high up in the northern part of the sky. Pegasus is also on view with its famous square practically due north.

SIGHTS OF INTEREST

In the southern hemisphere the night sky offers several objects that make good targets for even modest amateur equipment. Look south to find the constellation Tucana, the Toucan. Within the boundaries of this constellation you can see 47 Tucanae, or NGC 104, one of the best globular clusters in the night sky. With the naked eye it appears as a slightly fuzzy star. Near 47 Tucanae lies the galaxy known as the Small Magellanic Cloud, or SMC, which is a great target for a

The Circlet
This asterism is formed by a ring of seven stars that make up the head of one of the fish in the constellation Pisces.

small telescope or a pair of binoculars, and can also be seen with the naked eye.

A hop over the constellation Hydrus, or the Little Water Snake, takes you to the constellations Dorado and Mensa, where you will find the Large Magellanic Cloud, or LMC. It can be seen with the naked eye and is a fine sight through a telescope. Turning your attention to the northern sky, you can find the Andromeda Galaxy, M31, in the constellation Andromeda, as well as the spiral galaxy M33, visible through binoculars or a small telescope in the constellation Triangulum. The Andromeda Galaxy is the closest major galaxy to the Milky Way, and twice as large.

ERIDANUS				
Size ranking	Brightest star	Genitive	Abbreviation	Highest in sky at 10 pm
6	Alpha (α) Eridani, or Achernar, 0.5	Eridani	Eri	November–January

Eridanus, the River, winds its way across the night sky, starting near the feet of Orion. It then meanders across the sky towards Cetus before passing the constellations Horologium, the Clock, and Caelum, the Chisel. Its brightest star Achernar, Alpha (α) Eridani, has a magnitude 0.5, and marks the end of this celestial river. Eridanus has few clusters or nebulae within it but contains some interesting double stars, such as 32 Eridani and Theta (θ) Eridani.

MIDNIGHT

NEPTUNE

URANUS

OCTOBER
NORTHERN LATITUDES

OBSERVATION TIMES		
Date	Standard time	Daylight-saving time
15 September	Midnight	1 am
1 October	11 pm	Midnight
15 October	10 pm	11 pm
1 November	9 pm	10 pm
15 November	8 pm	9 pm

LOOKING **NORTH**

As the Summer Triangle (Vega, Deneb, and Altair) moves to the west the Milky Way arches overhead, and some winter constellations start peeking over the horizon. Look out for the Double Cluster in Perseus (see p.22) and the open clusters M36, M37, and M38 in Auriga, rising in the east. The Hyades and Pleiades star clusters in Taurus start coming back into view and the open clusters M52, NGC 457, and M103 high in the sky in Cassiopeia are also worth spotting. These are all good binocular targets.

The Auriga Clusters
Use binoculars to look out for the open clusters M36, M37, and M38. A telescope also shows the loose open cluster NGC 2281 nearby, which contains around 30 stars.

LOOKING **SOUTH**

While the beautiful Andromeda Galaxy, M31, is still on show this month, do not overlook another interesting galaxy close by. M33, or the Triangulum Galaxy is just above the constellation Aries, and can be glimpsed with the naked eye from a very dark sky site. A pair of binoculars or a small telescope will show this beautiful spiral galaxy's misty, oval form. If you are observing with a telescope, make sure to spot the lovely double star Gamma (γ) Arietis in the nearby constellation Aries.

The Andromeda Galaxy
Through a small telescope M31, in the constellation Andromeda, appears as a fuzzy grey ellipse with a brighter core. Larger apertures will help discern the galaxy's dark, dusty lanes.

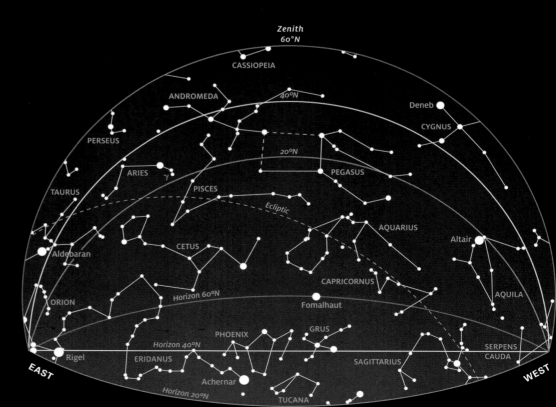

OCTOBER
SOUTHERN LATITUDES

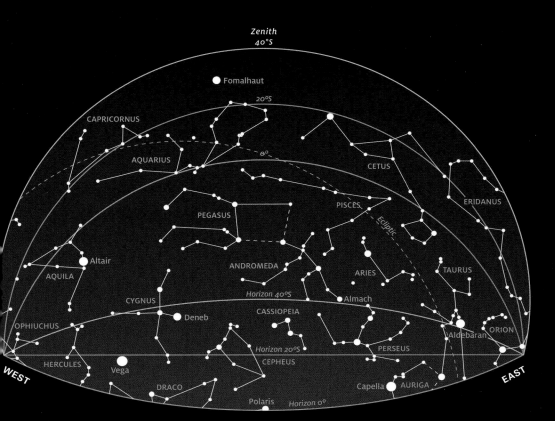

Zenith 40°S

Fomalhaut

20°S

CAPRICORNUS

AQUARIUS

0°

CETUS

PISCES

Ecliptic

ERIDANUS

PEGASUS

Altair

ANDROMEDA

ARIES

TAURUS

AQUILA

Horizon 40°S

Almach

CYGNUS

CASSIOPEIA

Aldebaran

ORION

Deneb

OPHIUCHUS

Horizon 20°S

PERSEUS

HERCULES

Vega

CEPHEUS

Capella

AURIGA

DRACO

Polaris *Horizon 0°*

WEST

EAST

Zenith 0°

AQUARIUS
20°S

Ecliptic

CETUS

Fomalhaut

40°S

PHOENIX

CAPRICORNUS

GRUS

AQUILA

ERIDANUS

Achernar

TUCANA

SAGITTARIUS

PAVO

OCTANS

ARA

Horizon 0°

PICTOR

Rigel

Canopus

SCORPIUS

Horizon 20°S

PUPPIS

CARINA

CENTAURUS

OPHIUCHUS

ORION

α Centauri

CANIS
MAJOR

β Centauri

Antares

EAST

WEST

VELA

LOOKING **NORTH**

The double star Almach, Gamma (γ) Andromedae in the constellation Andromeda, is a lovely target for a small telescope when looking north. One of the stars has an orange tint, while the other is a beautiful blue. In the same constellation, quite close to Almach, is the magnitude 5.5 open cluster NGC 752. Binoculars or small telescopes reveal it covering an area larger than the full Moon. Finally, be sure to track down M31 in Andromeda and M33 just above Aries.

NGC 752
Composed of around 70 stars, the loose open star cluster NGC 752 is a fine sight in a small telescope using a low magnification. It can be found north of Andromeda's star Almach in the east.

LOOKING **SOUTH**

Two of the celestial showpieces of the southern skies are on show this month: the Large and Small Magellanic Clouds. These galaxies lie relatively close in space to the Milky Way. The Small Magellanic Cloud (SMC) in Tucana can be seen with the naked eye, as can the Large Magellanic Cloud (LMC), which is a magnificent sight on the Dorado–Mensa border. Binoculars or small telescopes reveal many star clusters and patches of nebulosity within the LMC.

Small Magellanic Cloud (SMC)
The irregular galaxy known to astronomers as the SMC sits in the constellation Tucana. Visible to the naked eye, it stretches roughly seven times the Moon's apparent diameter across the sky.

OCTOBER | NORTHERN LATITUDES

STAR MAGNITUDES

DEEP-SKY OBJECTS

POINTS OF REFERENCE

LOOKING NORTH

OBSERVATION TIMES		
Date	**Standard time**	**Daylight-saving time**
15 September	Midnight	1 am
1 October	11 pm	Midnight
15 October	10 pm	11 pm
1 November	9 pm	10 pm
15 November	8 pm	9 pm

OCTOBER | NORTHERN LATITUDES

LOOKING SOUTH

STAR MOTION

North

South

STAR MAGNITUDES

• -1 • 0 • 1 • 2 · 3 · 4 · 5 ⊙ Variable star

DEEP-SKY OBJECTS

🌀 Galaxy ⭘ Globular cluster ⁂ Open cluster ✺ Diffuse nebula ⬭ Planetary nebula

POINTS OF REFERENCE

Horizons | 60°N | 40°N | 20°N | **Zeniths** | 60°N | 40°N | 20°N | Ecliptic

EAST

WEST

WEST

SOUTHWEST

SOUTH

SOUTHEAST

OPHIUCHUS

SCUTUM

M16

M25

M11

AQUILA

SAGITTA

Altair

M27

DELPHINUS

EQUULEUS

M15

SAGITTARIUS

M55

CAPRICORNUS

M2

M30

MICROSCOPIUM

PISCIS AUSTRINUS

Fomalhaut

INDUS

GRUS

TUCANA

Achernar

PEGASUS

ANDROMEDA

AQUARIUS

SCULPTOR

PHOENIX

HOROLOGIUM

PISCES

CETUS

Mira

FORNAX

ERIDANUS

TRIANGULUM

M33

ARIES

Ecliptic

TAURUS

ORION

OCTOBER | SOUTHERN LATITUDES

LOOKING NORTH

STAR MAGNITUDES

| ★ −1 | ★ 0 | • 1 | • 2 | • 3 | • 4 | • 5 | ⊙ Variable star |

DEEP-SKY OBJECTS

| 🌀 Galaxy | ⊛ Globular cluster | ❋ Open cluster | ✺ Diffuse nebula | ◯ Planetary nebula |

POINTS OF REFERENCE

Horizons	0°	20°S	40°S
Zeniths	+ 0°	+ 20°S	+ 40°S
			Ecliptic

OBSERVATION TIMES

Date	Standard time	Daylight-saving time
15 September	Midnight	1 am
1 October	11 pm	Midnight
15 October	10 pm	11 pm
1 November	9 pm	10 pm
15 November	8 pm	9 pm

OCTOBER | SOUTHERN LATITUDES

LOOKING SOUTH

STAR MOTION

North

South

STAR MAGNITUDES

−1 0 1 2 3 4 5 ● Variable star

DEEP-SKY OBJECTS

Galaxy Globular cluster Open cluster Diffuse nebula Planetary nebula

POINTS OF REFERENCE

Horizons | 0° 20°S 40°S
Zeniths | 0° 20°S 40°S Ecliptic

NOVEMBER

We are now deep into the wonderful dark nights of autumn in the northern hemisphere, where the constellations Orion, Taurus, Auriga, and Gemini are on show. In the southern hemisphere Cetus, Eridanus, and Aquarius are high in the sky.

NORTHERN LATITUDES

THE STARS

Two interesting constellations, Perseus and Cassiopeia, are almost overhead this month. Looking south, the constellations Pegasus and Andromeda are sitting high in the sky. In the east, the grand constellations Orion, Taurus, and Auriga are also visible.

SIGHTS OF INTEREST

In the northern hemisphere November's skies contain the lingering sights of summer, including M31 and M33, as well as some new objects. The open clusters NGC 457 and NGC 663 in the "W" shaped constellation Cassiopeia make excellent binocular targets. A small telescope shows the glittering pair of clusters in Perseus listed as NGC 869 and NGC 884, collectively known as the Double Cluster. There are also several fine open clusters on show in Auriga.

METEOR SHOWERS

Look out for the peak of the Taurid meteor shower during the first week of November. If the skies are clear and dark you may see 10 meteors an hour, coming from a point south of M45. Another meteor shower, the Leonids, peaks around 17 November. At its peak you can observe up to 10 meteors an hour, shooting from the direction of Leo's head.

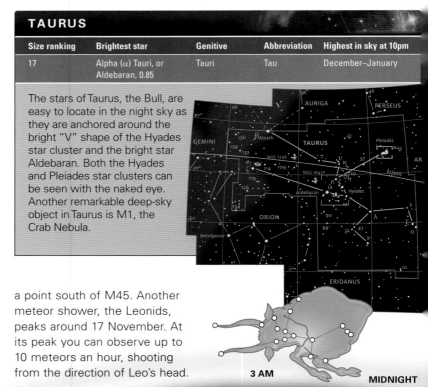

TAURUS

Size ranking	Brightest star	Genitive	Abbreviation	Highest in sky at 10pm
17	Alpha (α) Tauri, or Aldebaran, 0.85	Tauri	Tau	December–January

The stars of Taurus, the Bull, are easy to locate in the night sky as they are anchored around the bright "V" shape of the Hyades star cluster and the bright star Aldebaran. Both the Hyades and Pleiades star clusters can be seen with the naked eye. Another remarkable deep-sky object in Taurus is M1, the Crab Nebula.

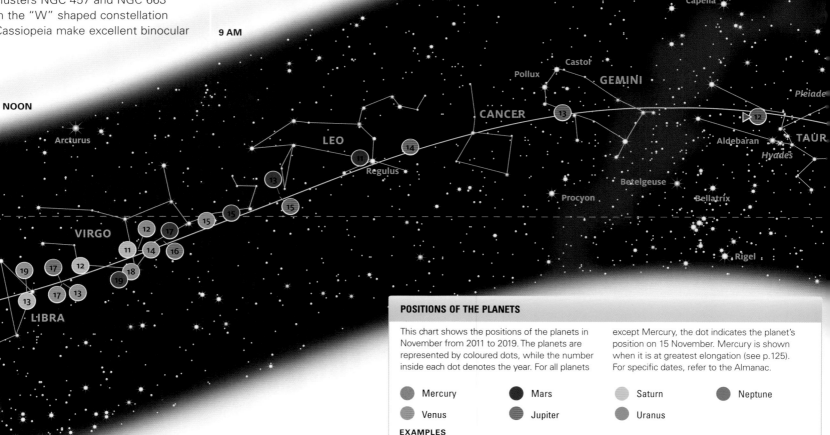

POSITIONS OF THE PLANETS

This chart shows the positions of the planets in November from 2011 to 2019. The planets are represented by coloured dots, while the number inside each dot denotes the year. For all planets except Mercury, the dot indicates the planet's position on 15 November. Mercury is shown when it is at greatest elongation (see p.125). For specific dates, refer to the Almanac.

- ⬤ Mercury
- ⬤ Venus
- ⬤ Mars
- ⬤ Jupiter
- ⬤ Saturn
- ⬤ Uranus
- ⬤ Neptune

EXAMPLES

⑪ Mars's position on 15 November 2011

◁⑪ Saturn's position on 15 November 2011. The arrow indicates that the planet is in retrograde motion (see p.125)

SOUTHERN LATITUDES

THE STARS

The constellations Eridanus and Cetus sit right above you this month. Eridanus, the River, is naturally long and winding and its end is marked by the bright star Achernar, which can be seen high in the sky almost due south. The constellation Phoenix sits close to Eridanus and below it, towards the direction of the south celestial pole, are the constellations Reticulum, the Net; Hydrus, the Little Water Snake; Tucana, the Toucan; and Octans, the Octant.

In the east you can locate Canis Major, which is hard to miss as it is home to the blazing star Sirius. Also coming into view in the east are

Orion and Taurus. It is easy to identify Orion, as it contains the bright stars Betelgeuse, Alpha (α) Orionis, and Rigel, Beta (β) Orionis. Look north to find the constellations Andromeda, Pisces, and Aries.

SIGHTS OF INTEREST

As Cetus is high in the sky, a large telescope will show you the interesting spiral galaxy M77, sitting very close to the star Delta (δ) Ceti. The beautiful Magellanic Clouds should be your next target. The Large Magellanic Cloud, or the LMC, sits across the border between the constellations Dorado, the Goldfish (or Swordfish), and Mensa, the Table Mountain. A small telescope is all you need to explore the sparkling star clusters as well as the Tarantula Nebula, or NGC 2070, nestled within the LMC. Meanwhile, a short distance away in the constellation

The variable star Mira

Omicron (o) Ceti, more popularly known as Mira, is a variable star in the constellation Cetus. Its brightness changes over time as it pulsates.

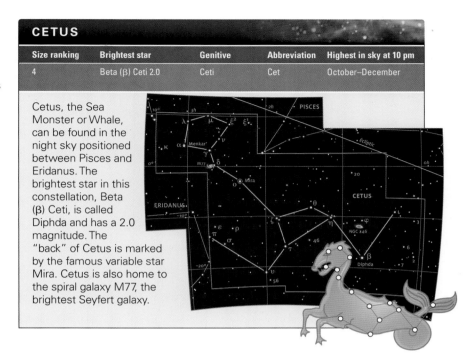

CETUS				
Size ranking	Brightest star	Genitive	Abbreviation	Highest in sky at 10 pm
4	Beta (β) Ceti 2.0	Ceti	Cet	October–December

Cetus, the Sea Monster or Whale, can be found in the night sky positioned between Pisces and Eridanus. The brightest star in this constellation, Beta (β) Ceti, is called Diphda and has a 2.0 magnitude. The "back" of Cetus is marked by the famous variable star Mira. Cetus is also home to the spiral galaxy M77, the brightest Seyfert galaxy.

Tucana, you can see the Small Magellanic Cloud and the globular cluster NGC 104, or 47 Tucanae. These are wonderful binocular or small telescope targets. Both the Large and Small Magellanic Cloud can be seen with the naked eye.

Looking towards the northeast, the Hyades and Pleiades open star clusters make excellent binocular targets. Also look out for the stunning spiral galaxy M33 and the glowing ellipse of Andromeda Galaxy, or M31, through a telescope.

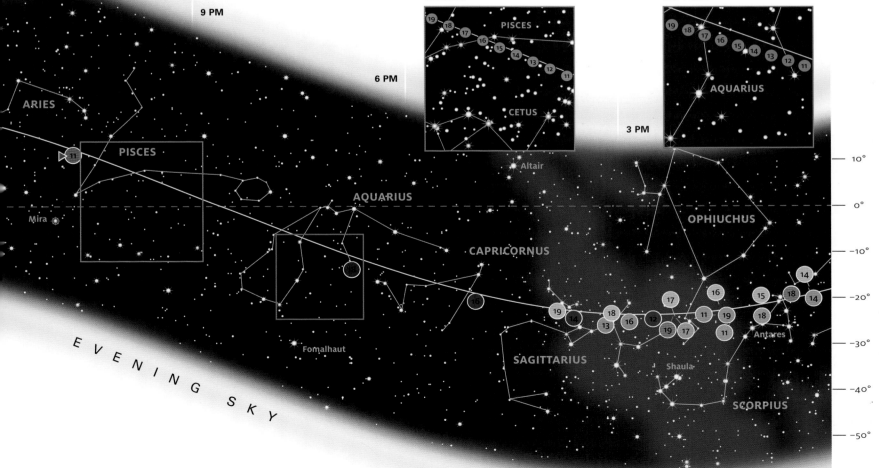

NOVEMBER
NORTHERN LATITUDES

LOOKING NORTH

Be sure to savour the sights of Cygnus, the Swan, before the constellation starts to sink below the horizon. A small telescope will show its beautiful double stars Omicron-1 (o¹) Cygni, 61 Cygni, and Albireo (see p.62), which marks the beak of the swan. A pair of binoculars will show the open clusters M29 and M39.

Also look out for the magnitude 7.3 open cluster M52, sitting high in the sky in Cassiopeia; the open clusters M36, M37, and M38 in Auriga (see p.46); and M35 in Gemini.

M29 in Cygnus
This open cluster can be found sitting against the background star fields of the Milky Way. Located a little way from the star Sadr, it is best observed with a small telescope.

LOOKING SOUTH

The Hyades (see p.23) and Pleiades (see p.38) star clusters in the constellation Taurus make a welcome return to winter skies in the east. A pair of binoculars is all you will need to explore these two open clusters. Both the Pleiades and Hyades are clearly visible to the naked eye, but binoculars will reveal the many glittering stars within them.

Other objects of interest to look out for include the Andromeda Galaxy, M31, and the Triangulum Galaxy, M33, sitting high in the sky.

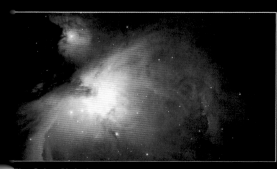

The Orion Nebula
Also known as M42, this is one of the finest nebulae in the night sky. A wonderful sight in all types of equipment, a small telescope shows its glowing cavernous gas clouds and embedded stars.

OBSERVATION TIMES		
Date	Standard time	Daylight-saving time
15 October	Midnight	1 am
1 November	11 pm	Midnight
15 November	10 pm	11 pm
1 December	9 pm	10 pm
15 December	8 pm	9 pm

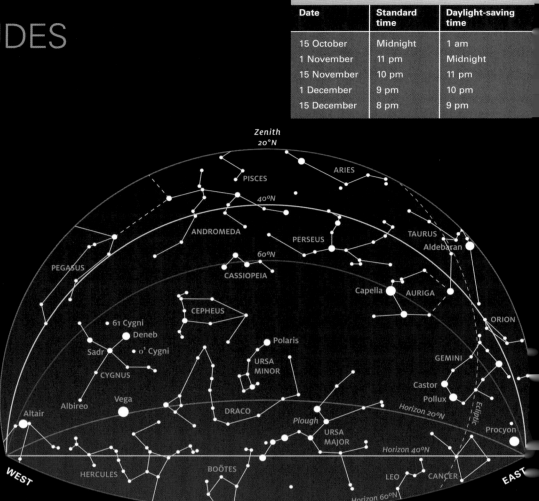

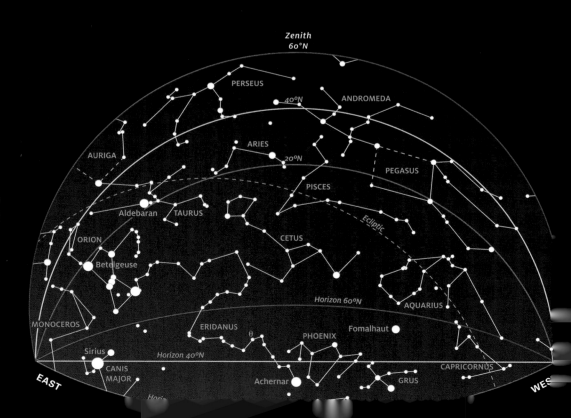

STAR MAGNITUDES

-1 0 1 2 • 3 and above

NOVEMBER
SOUTHERN LATITUDES

Zenith
40°S

20°S

ERIDANUS

Fomalhaut

CETUS

AQUARIUS

0°

Ecliptic

o² Eridani

PISCES

TAURUS

ARIES

Rigel

PEGASUS

ORION

Aldebaran

MONOCEROS

ANDROMEDA

Betelgeuse

Horizon 40°S

CASSIOPEIA

PERSEUS

Capella

AQUILA

Horizon 20°S

AURIGA

GEMINI

WEST

Altair

Procyon

CYGNUS

Deneb

EAST

Horizon 0°

CEPHEUS

Castor

Polaris

LOOKING **NORTH**

Look out for the globular cluster M79 lying under the feet of Orion, the Hunter, in the east. This cluster is visible through a small telescope. In the constellation Eridanus above Orion, you will find an interesting multiple star system, Omicron-2 (o²) Eridani with three components, and a double star Theta (θ) Eridani. Both are visible with small telescopes. Deep-sky observers with dark skies will be able to spot the galaxy NGC 1300 in the same constellation using large telescopes

NGC 1300

The barred spiral galaxy NGC 1300 is located about 69 million light-years from Earth. It is particularly faint and can be seen with a very large amateur telescope

LOOKING **SOUTH**

There is a lot to see in the southern skies this month with just the naked eye. Start off by looking for the Large Magellanic Cloud (LMC), west of the constellation Pictor, and the Small Magellanic Cloud (SMC), in Tucana. These are both irregular galaxies close to the Milky Way. The globular cluster 47 Tucanae can be seen with the naked eye as a hazy star very close to the Small Magellanic Cloud. The open clusters NGC 2362 and M41 in Canis Major in the east also make for good small-telescope targets

Zenith
0°

CETUS

20°S

Ecliptic

40°S

PHOENIX

Fomalhaut

ERIDANUS

Achernar

GRUS

AQUARIUS

Rigel

TUCANA

ORION

CAPRICORNUS

CANIS
MAJOR

Canopus

PICTOR

OCTANS

PAVO

Sirius

CARINA

SAGITTARIUS

MONOCEROS

PUPPIS

Horizon 0°

ARA

Horizon 20°S

VELA

Horizon 40°S

47 Tucanae

This globular cluster in Tucana is a wonderful sight through a

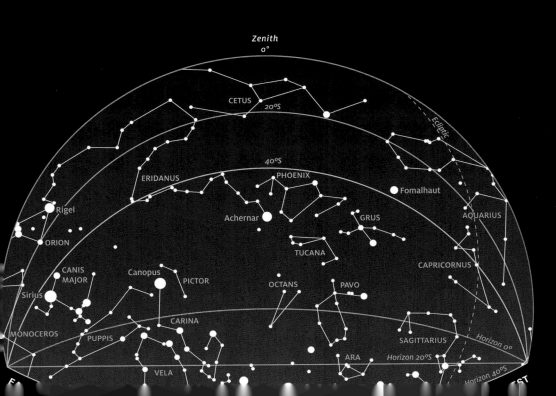

STAR MAGNITUDES

DEEP-SKY OBJECTS

POINTS OF REFERENCE

NOVEMBER | NORTHERN LATITUDES

LOOKING NORTH

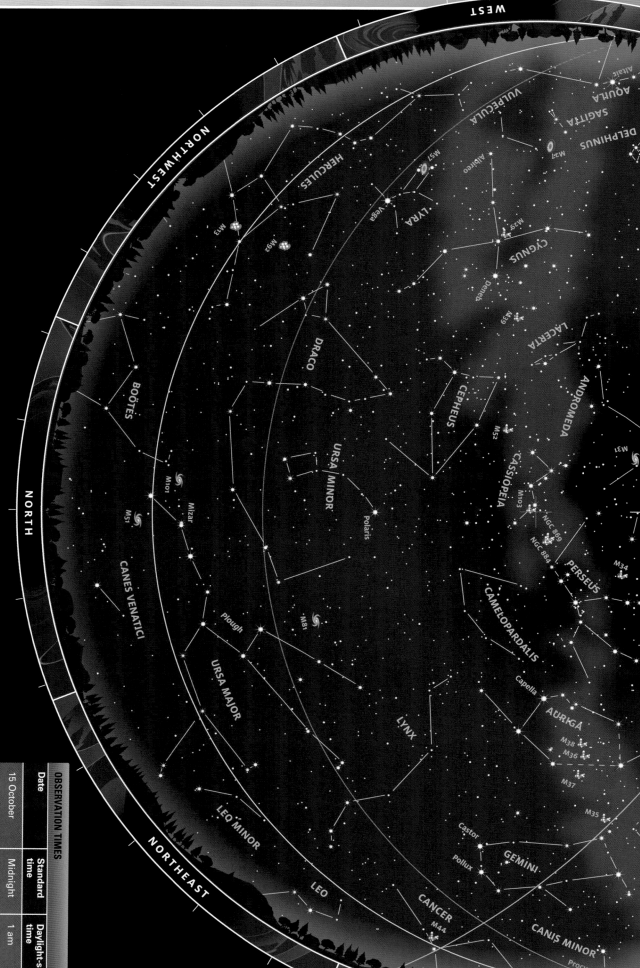

OBSERVATION TIMES		
Date	Standard time	Daylight-saving time
15 October	Midnight	1 am
1 November	11 pm	Midnight
15 November	10 pm	11 pm
1 December	9 pm	10 pm
15 December	8 pm	9 pm

NOVEMBER | NORTHERN LATITUDES

LOOKING SOUTH

WEST

EAST

SOUTHEAST

SOUTH

SOUTHWEST

WEST

STAR MOTION

North

South

STAR MAGNITUDES

★ -1 ● 0 ● 1 ● 2 · 3 · 4 · 5 ⊛ Variable star

DEEP-SKY OBJECTS

🌀 Galaxy ● Globular cluster ✦ Open cluster ✿ Diffuse nebula ⬮ Planetary nebula

POINTS OF REFERENCE

Horizons | 60°N | 40°N | 20°N | Zeniths | 60°N | 40°N | 20°N | Ecliptic

AQUILA
CAPRICORNUS
EQUULEUS
M72
M15
PEGASUS
ANDROMEDA
AQUARIUS
M30
MICROSCOPIUM
PISCIS AUSTRINUS
Fomalhaut
GRUS
SCULPTOR
TUCANA
PHOENIX
Achernar
PISCES
TRIANGULUM
M33
ARIES
CETUS
Mira
FORNAX
ERIDANUS
HOROLOGIUM
RETICULUM
DORADO
CAELUM
Ecliptic
M45 (Pleiades)
TAURUS
Hyades
Aldebaran
M1
ORION
Bellatrix
Rigel
Betelgeuse
M42
LEPUS
COLUMBA
MONOCEROS
CANIS MAJOR
M50
Sirius
M41

NOVEMBER | SOUTHERN LATITUDES

LOOKING NORTH

OBSERVATION TIMES		
Date	**Standard time**	**Daylight-saving time**
15 October	Midnight	1 am
1 November	11 pm	Midnight
15 November	10 pm	11 pm
1 December	9 pm	10 pm
15 December	8 pm	9 pm

NOVEMBER | SOUTHERN LATITUDES

LOOKING SOUTH

WEST

SOUTHWEST

SOUTH

SOUTHEAST

EAST

STAR MOTION

North

South

POINTS OF REFERENCE

| Horizons | 0° | 20°S | 40°S | | Zeniths | + 0° | + 20°S | + 40°S | | Ecliptic |

DEEP-SKY OBJECTS

Galaxy | Globular cluster | Open cluster | Diffuse nebula | Planetary nebula

STAR MAGNITUDES

-1 | 0 | 1 | 2 | 3 | 4 | 5 | Variable star

CAPRICORNUS
AQUARIUS
PISCIS AUSTRINUS
M30
Fomalhaut
SCULPTOR
MICROSCOPIUM
M55
M54 M69
M70
SAGITTARIUS
M7
M17
CORONA AUSTRALIS
SCORPIUS
Shaula
TELESCOPIUM
ARA
NORMA
GRUS
INDUS
PAVO
CETUS
FORNAX
PHOENIX
NGC 104
TUCANA
SMC
OCTANS
APUS
TRIANGULUM AUSTRALE
CIRCINUS
β Centauri
α Centauri
Achernar
HYDRUS
HOROLOGIUM
ERIDANUS
RETICULUM
MENSA
LMC
CHAMAELEON
MUSCA
Acrux
CRUX
Becrux
Gacrux
CENTAURUS
CAELUM
DORADO
VOLANS
NGC 1300
PICTOR
Canopus
CARINA
COLUMBA
PUPPIS
VELA
LEPUS
M79
Adhara
Sirius
CANIS MAJOR
NGC 2362
PYXIS
M50
M47 M93
MONOCEROS
M46 M41

DECEMBER

As we round off the year, the northern skies contain the spectacular constellations Orion, Taurus, Gemini, and Auriga. These can also be glimpsed from the southern hemisphere, along with Vela and Carina.

NORTHERN LATITUDES

THE STARS

Looking north, the constellations Perseus, Auriga, and Andromeda can be found sitting high in the sky. In the southeast, it is impossible to ignore the magnificent sight of Orion, the Hunter. Orion leads the winter constellations, which include Taurus in the south and Gemini in the east. The Winter Triangle formed by Betelgeuse, Sirius, and Procyon is easy to locate in the southeast.

SIGHTS OF INTEREST

Observers in the northern hemisphere are spoilt for choice this month. In Orion lies the beautiful nebula M42, which is a superb target for a small telescope or binoculars. In Taurus, the Bull, you can locate two striking star clusters. The Hyades is a large cluster outlining the bull's face and the Pleiades is possibly the finest open cluster in the sky. Auriga too has several stunning open clusters to look out for.

The Geminids
If you are observing the Geminids, you might spot a very bright meteor. These "fireballs" add to the excitement of watching these celestial fireworks.

ORION				
Size ranking	Brightest star	Genitive	Abbreviation	Highest in sky at 10 pm
26	Beta (β) Orionis, or Rigel, 0.2	Orionis	Ori	December–January

Orion, the Hunter, is one of the greatest constellations in the whole night sky. You can find it by spotting the prominent line of three stars that form the hunter's "belt," as well as its distinctly coloured stars, Rigel and Betelgeuse. Orion also contains one of the most stunning nebulae in the sky, M42, also known as the Orion Nebula.

METEOR SHOWER

The Geminid meteor shower peaks around 13–14 December. At its peak you can expect to see around one meteor per minute streaking across the sky from the direction of Gemini.

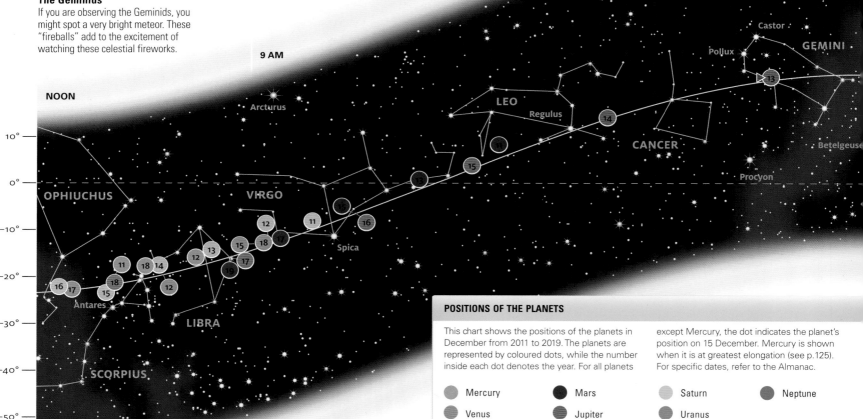

POSITIONS OF THE PLANETS

This chart shows the positions of the planets in December from 2011 to 2019. The planets are represented by coloured dots, while the number inside each dot denotes the year. For all planets except Mercury, the dot indicates the planet's position on 15 December. Mercury is shown when it is at greatest elongation (see p.125). For specific dates, refer to the Almanac.

- Mercury
- Venus
- Mars
- Jupiter
- Saturn
- Uranus
- Neptune

EXAMPLES

⊙11 Mars' position on 15 December 2011

▷⊙11 Saturn's position on 15 December 2011. The arrow indicates that the planet is in retrograde motion (see p.125)

SOUTHERN LATITUDES

THE STARS

The constellations Taurus, Gemini, Orion, and Auriga are visible from the southern hemisphere this month. Looking towards the north you can spot the distinct "V" shape of the Hyades star cluster, which marks the head of the constellation Taurus. Just next to it towards the northeast is Orion, a magnificent sight with its bright stars Rigel and Betelgeuse. Emerging from the foot of Orion, Eridanus meanders high across the sky. Look out for Perseus and Auriga below Taurus in the

north. Auriga can be found by locating the bright star Capella, which is low in the sky this month. If you look roughly northeast Gemini is also on show, sitting low in the sky near the horizon. Meanwhile Canis Major, Vela, and Carina can be located in the southeast. December is also a great time to look out for the Magellanic Clouds in the south.

SIGHTS OF INTEREST

The Magellanic Clouds can be seen sitting in the southern part of the night sky this month. The Small Magellanic Cloud sits in Tucana, while the Large Magellanic Cloud hovers on the border of the constellations Dorado, the

Goldfish, and Mensa, the Table Mountain. Look out for the Tarantula Nebula, or NGC 2070, in the Large Magellanic Cloud. High in the northeast lies the fantastic nebula M42, in Orion. In the nearby constellation

Taurus, you have a great opportunity to observe two open star clusters, the Hyades and the Pleiades. The Pleiades star cluster, or M45, can be seen with the naked eye and is a wonderful sight in a small telescope.

The Large Magellanic Cloud
You can see the Large Magellanic Cloud with the naked eye. A small telescope reveals the Tarantula Nebula, NGC 2070, embedded within it.

LEPUS

Size ranking	Brightest star	Genitive	Abbreviation	Highest in sky at 10 pm
51	Alpha (α) Leporis, or Arneb, 2.6	Leporis	Lep	January

The constellation Lepus, the Hare, sits right under the feet of the magnificent Orion, between Canis Major and Eridanus. Unlike its neighbours, Canis Major and Orion, Lepus contains few bright stars. Even so, it does have a handful of targets for you to observe. These include M79, a magnitude 8 globular cluster, and a small group of stars catalogued as NGC 2017, both of which can be seen with a small telescope.

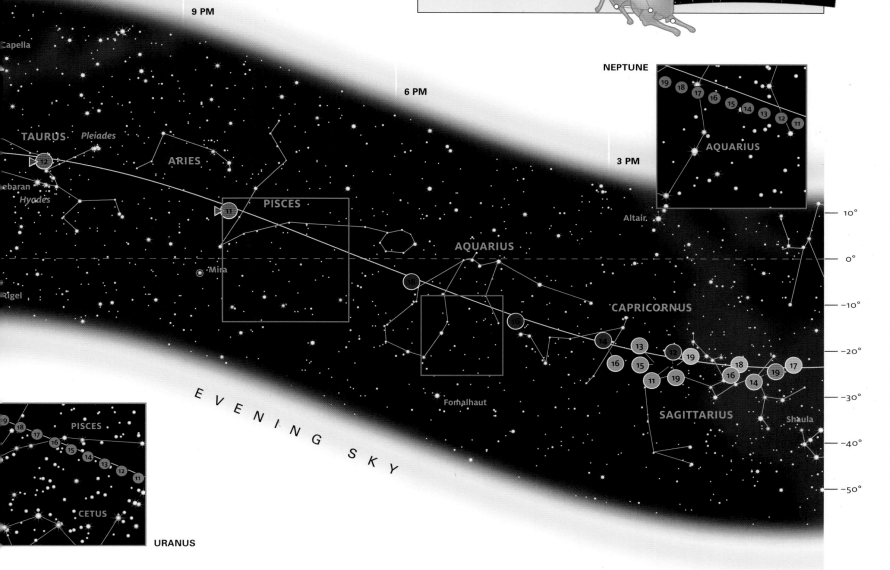

DECEMBER
NORTHERN LATITUDES

OBSERVATION TIMES		
Date	Standard time	Daylight-saving time
15 November	Midnight	1 am
1 December	11 pm	Midnight
15 December	10 pm	11 pm
1 January	9 pm	10 pm
15 January	8 pm	9 pm

LOOKING NORTH

There are several interesting sights on show in the constellation Gemini, the Twins, in the east. Castor, Gemini's second brightest star, is an interesting multiple star, while the open cluster M35, sitting close to the feet of one of the twins, is a great target for binoculars or a small telescope.

Other December treats include the Double Cluster (see p.22) in Perseus; the Andromeda Galaxy (see p.94), which is still high in the sky; and the Milky Way rising up through Cygnus.

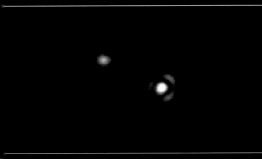

Castor
A multiple star system, Castor, or Alpha (α) Geminorum, can be seen with a small telescope. The two main stars orbit each other roughly once about every 468 years.

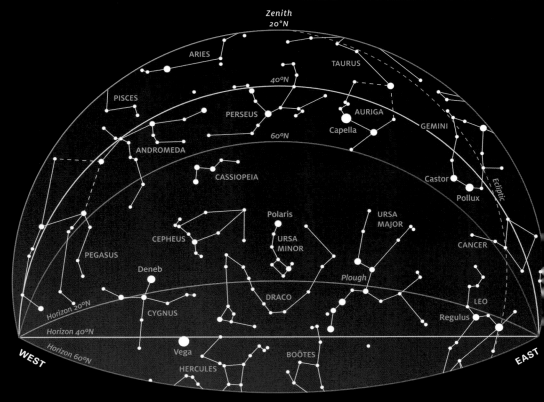

LOOKING SOUTH

The Orion Nebula is undoubtedly the prime target in this part of the northern skies (see p.102). It can be found in Orion's Sword, which drops down from the middle of the "belt" stars. Through a small telescope, you should be able to see the nebula, resembling a celestial cave, with stars embedded in its heart. Larger apertures reveal the nebula's swirls of gas, which make this one of the finest deep-sky objects in the sky. Also on show are the Hyades and Pleiades – two star clusters in Taurus.

Orion
The constellation Orion, the Hunter, is a magnificent sight in the winter night sky. The star Betelgeuse marks the shoulder of the Hunter, while Rigel marks his f...

STAR MAGNITUDES

● -1　● 0　● 1　● 2　• 3 and above

DECEMBER
SOUTHERN LATITUDES

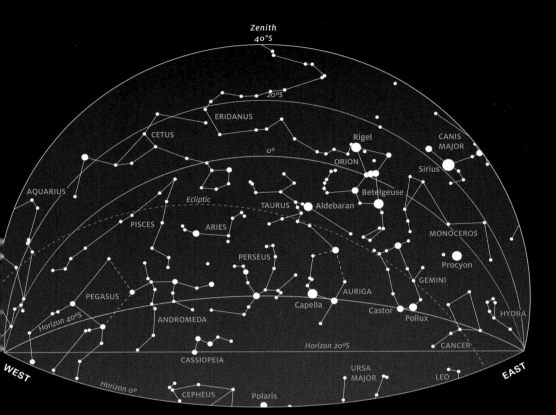

Zenith
40°S

WEST

EAST

LOOKING NORTH

The constellation Orion has many interesting objects in and around it to entice observers in the southern hemisphere. Sky gazers should look out for the Orion Nebula (see p.102) which is a fine target for binoculars or a small telescope. The Pleiades and Hyades star clusters can be enjoyed with the naked eye. Meanwhile, in the nearby constellations Auriga, Monoceros, and Puppis, there are several open clusters on show, such as M36, M37, M38, M50, M46, and M47.

M38 in Auriga
The magnitude 6.4 open cluster M38 is the most scattered of the three famous Messier clusters in Auriga in the north at the moment. It lies 4,200 light-years from Earth.

LOOKING SOUTH

If you are observing with binoculars, there is a great deal to see looking south. NGC 3114 and NGC 2516 are both prominent open clusters worth observing in Carina in the southeast. The Eta (ε) Carinae Nebula, or NGC 3372, is a bright diffuse nebula visible through binoculars or a small telescope. The bright open cluster IC 2602, or the Southern Pleiades, is a great binocular object. Also look out for the Small Magellanic Cloud in Tucana and the Large Magellanic Cloud, just west of Pictor.

Zenith
0°

EAST

WEST

The Large Magellanic Cloud (LMC)
A conspicuous naked-eye object, the LMC sits on the border of the constellations Mensa and Dorado. A small telescope will show star clusters and bright patches of nebulosity within it.

DECEMBER | NORTHERN LATITUDES

LOOKING NORTH

OBSERVATION TIMES		
Date	Standard time	Daylight-saving time
15 November	Midnight	1 am
1 December	11 pm	Midnight
15 December	10 pm	11 pm
1 January	9 pm	10 pm
15 January	8 pm	9 pm

DECEMBER | NORTHERN LATITUDES

LOOKING SOUTH

STAR MOTION

North

South

STAR MAGNITUDES

* -1   * 0   · 1   · 2   • 3   · 4   · 5   ⊛ Variable star

DEEP-SKY OBJECTS

🌀 Galaxy   ⊙ Globular cluster   ✳ Open cluster   ✿ Diffuse nebula   ◉ Planetary nebula

POINTS OF REFERENCE

Horizons | 60°N | 40°N | 20°N | Zeniths | 60°N | 40°N | 20°N | Ecliptic

DECEMBER | SOUTHERN LATITUDES

LOOKING NORTH

OBSERVATION TIMES		
Date	**Standard time**	**Daylight-saving time**
15 November	Midnight	1 am
1 December	11 pm	Midnight
15 December	10 pm	11 pm
1 January	9 pm	10 pm
15 January	8 pm	9 pm

DECEMBER | SOUTHERN LATITUDES

LOOKING SOUTH

STAR MOTION

North

South

STAR MAGNITUDES

-1 0 1 2 3 4 5 ⊛ Variable star

DEEP-SKY OBJECTS

🌀 Galaxy ✦ Globular cluster ✧ Open cluster 🦋 Diffuse nebula ⬭ Planetary nebula

POINTS OF REFERENCE

Horizons | 0° | 20°S | 40°S

Zeniths | 0° | 20°S | 40°S

| Ecliptic

WEST

WEST

SOUTHWEST

SOUTH

SOUTHEAST

EAST

Constellations and objects labeled:

AQUARIUS, PISCIS AUSTRINUS, Fomalhaut, CAPRICORNUS, M30, SAGITTARIUS, MICROSCOPIUM, GRUS, INDUS, SCULPTOR, CETUS, TELESCOPIUM, PAVO, ARA, TRIANGULUM AUSTRALE, OCTANS, TUCANA, NGC 104, SMC, HYDRUS, Achernar, PHOENIX, FORNAX, HOROLOGIUM, RETICULUM, MENSA, APUS, CHAMAELEON, CIRCINUS, β Centauri, α Centauri, ERIDANUS, CAELUM, LMC, DORADO, VOLANS, MUSCA, Acrux, Becrux, CRUX, Gacrux, CENTAURUS, LEPUS, COLUMBA, Canopus, PICTOR, CARINA, NGC 2516, NGC 3372, IC 2602, NGC 3114, CANIS MAJOR, Sirius, M41, Adhara, PUPPIS, VELA, M47, M46, PYXIS, ANTLIA, HYDRA

ALMANAC

This section contains astronomical calendars listing major celestial events for the years 2011–2019. These calendars show the phases of the Moon, eclipses of the Sun and the Moon, and motions of the planets. The latter include the greatest western and eastern elongations (the angle between the Sun and a planet) of Mercury and Venus. Also shown are the oppositions of Mars, Jupiter, and Saturn, when these planets are on the opposite side of Earth from the Sun and visible in the sky through the night.

2011

This year sees an unusually large number of eclipses – four partial solar eclipses and two total lunar eclipses. Other highlights include a close conjunction of Jupiter and Uranus, which began in 2010 and continues through January 201 A conjunction is a close alignment of two bodies in the sky and occurs when both planets lie in the same line of sight as viewed from Earth.

| | Full Moon | | New Moon | | Total eclipse of the Moon | | Partial eclipse of the Moon | | Partial eclipse of the Sun | | Annular eclipse of the Sun | | Total eclipse of the Sun |

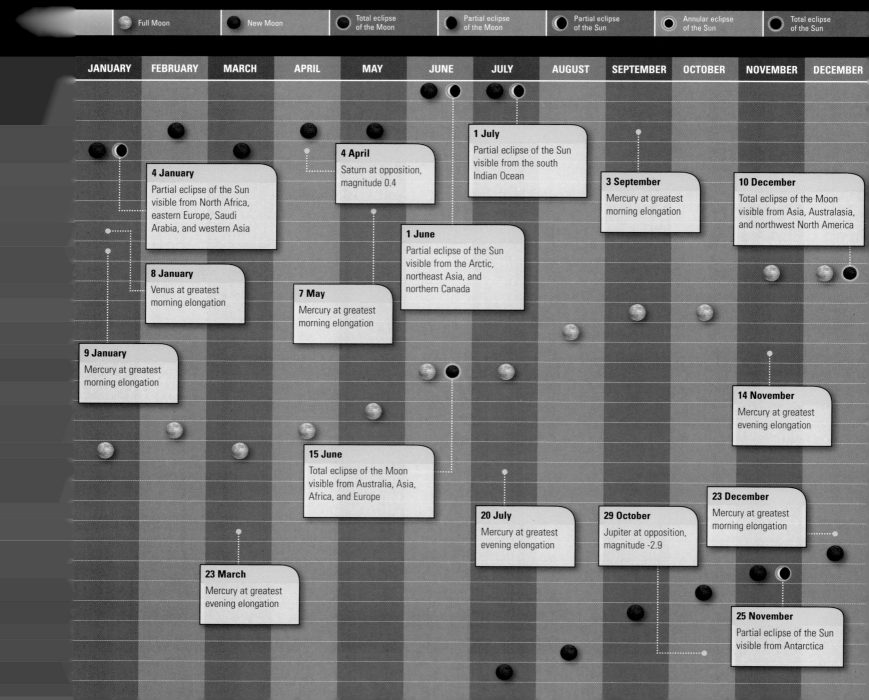

JANUARY FEBRUARY MARCH APRIL MAY JUNE JULY AUGUST SEPTEMBER OCTOBER NOVEMBER DECEMBER

4 January
Partial eclipse of the Sun visible from North Africa, eastern Europe, Saudi Arabia, and western Asia

8 January
Venus at greatest morning elongation

9 January
Mercury at greatest morning elongation

4 April
Saturn at opposition, magnitude 0.4

7 May
Mercury at greatest morning elongation

1 June
Partial eclipse of the Sun visible from the Arctic, northeast Asia, and northern Canada

15 June
Total eclipse of the Moon visible from Australia, Asia, Africa, and Europe

1 July
Partial eclipse of the Sun visible from the south Indian Ocean

20 July
Mercury at greatest evening elongation

23 March
Mercury at greatest evening elongation

3 September
Mercury at greatest morning elongation

29 October
Jupiter at opposition, magnitude -2.9

10 December
Total eclipse of the Moon visible from Asia, Australasia, and northwest North America

14 November
Mercury at greatest evening elongation

23 December
Mercury at greatest morning elongation

25 November
Partial eclipse of the Sun visible from Antarctica

Full Moon	New Moon	Total eclipse of the Moon	Partial eclipse of the Moon	Partial eclipse of the Sun	Annular eclipse of the Sun	Total eclipse of the Sun

JANUARY	FEBRUARY	MARCH	APRIL	MAY	JUNE	JULY	AUGUST	SEPTEMBER	OCTOBER	NOVEMBER	DECEMBER

3 March
Mars at opposition, magnitude -1.2

5 March
Mercury at greatest evening elongation

1 July
Mercury at greatest evening elongation

3 December
Jupiter at opposition, magnitude -2.8

4 December
Mercury at greatest morning elongation

4 June
Partial eclipse of the Moon visible from western regions of North and South America, Pacific Ocean, Australasia, and east Asia

5–6 June
Transit of Venus visible from east Asia, Australasia, and northwestern North America

15 April
Saturn at opposition, magnitude 0.2

15 August
Venus at greatest morning elongation

18 April
Mercury at greatest morning elongation

13–14 November
Total eclipse of the Sun visible from south Pacific Ocean. Partial eclipse visible from eastern Australia and New Zealand

16 August
Mercury at greatest morning elongation

20–21 May
Annular eclipse of the Sun visible from Japan and north Pacific Ocean. Partial eclipse visible from east Asia and western North America

26 October
Mercury at greatest evening elongation

27 March
Venus at greatest evening elongation

2012

...dition to a partial eclipse
...e Moon, this year sees a
...and an annular eclipse of
...un. However, the highlight
...12 is the rare transit of Venus.
...ext such event will not occur
...he year 2117.

...t of Venus
...June Venus will pass across the face
...Sun in a rare transit event that reveals

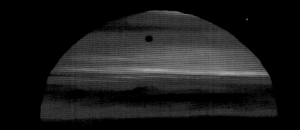

Annular eclipse
In May the Moon will lie near its furthest point from Earth as it eclipses the Sun. Even when perfectly aligned, a thin ring of sunlight will remain visible.

2013

annular eclipses of the Sun
visible this year, of which one
ars total from some locations.
rtial lunar eclipse is widely
le, and there is a rare
unction of the inferior
ets Mercury and Venus.

spread eclipse
eclipses of the Sun can be seen over
areas than total eclipses. The November
eclipse will be visible from many regions.

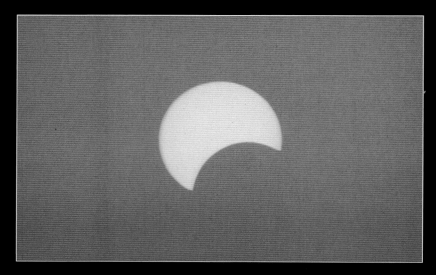

Mercury and Venus
In June Mercury and Venus will make a rare
close approach in evening skies, tracing paths
similar to those seen in this time-lapse image.

	Full Moon	New Moon	Total eclipse of the Moon	Partial eclipse of the Moon	Partial eclipse of the Sun	Annular eclipse of the Sun	Total eclipse of the Sun

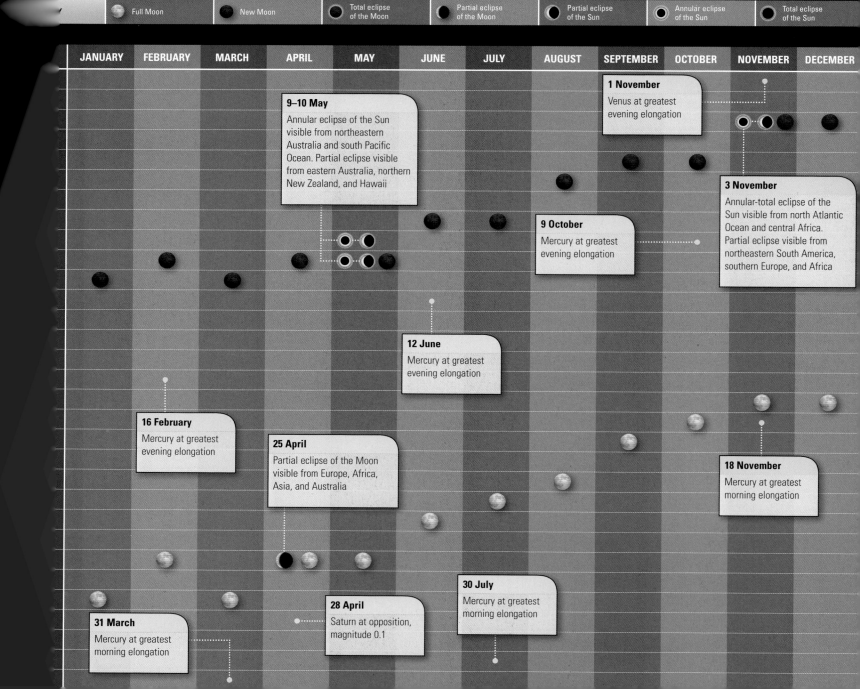

JANUARY	FEBRUARY	MARCH	APRIL	MAY	JUNE	JULY	AUGUST	SEPTEMBER	OCTOBER	NOVEMBER	DECEMBER

1 November
Venus at greatest
evening elongation

9–10 May
Annular eclipse of the Sun
visible from northeastern
Australia and south Pacific
Ocean. Partial eclipse visible
from eastern Australia, northern
New Zealand, and Hawaii

3 November
Annular-total eclipse of the
Sun visible from north Atlantic
Ocean and central Africa.
Partial eclipse visible from
northeastern South America,
southern Europe, and Africa

9 October
Mercury at greatest
evening elongation

12 June
Mercury at greatest
evening elongation

16 February
Mercury at greatest
evening elongation

25 April
Partial eclipse of the Moon
visible from Europe, Africa,
Asia, and Australia

18 November
Mercury at greatest
morning elongation

30 July
Mercury at greatest
morning elongation

28 April
Saturn at opposition,
magnitude 0.1

31 March
Mercury at greatest
morning elongation

KEY

Full Moon	New Moon	Total eclipse of the Moon	Partial eclipse of the Moon
Partial eclipse of the Sun	Annular eclipse of the Sun	Total eclipse of the Sun	

	JANUARY	FEBRUARY	MARCH	APRIL	MAY	JUNE	JULY	AUGUST	SEPTEMBER	OCTOBER	NOVEMBER	DECEMBER
1												
2												
3												
4												
5												
6												
7												
8												
9												
10												
11												
12												
13												
14												
15												
16												
17												
18												
19												
20												
21												
22												
23												
24												
25												
26												
27												
28												
29												
30												
31												

1 November
Mercury at greatest morning elongation

5 January
Jupiter at opposition, magnitude -2.7

8 April
Mars at opposition, magnitude -1.5

10 May
Saturn at opposition, magnitude 0.1

14 March
Mercury at greatest morning elongation

8 October
Total eclipse of the Moon visible from North America, Australasia, and east Asia

12 July
Mercury at greatest morning elongation

15 April
Total eclipse of the Moon visible from North America, South America, and New Zealand

21 September
Mercury at greatest evening elongation

22 March
Venus at greatest morning elongation

25 May
Mercury at greatest evening elongation

29 April
Partial eclipse of the Sun visible from west Australia

31 January
Mercury at greatest evening elongation

28 October
Partial eclipse of the Sun visible from western North America

2014

In addition to a pair of partial solar eclipses and two total lunar eclipses, this year sees rare celestial events such as a brief occultation (see p.125) of a bright star by an asteroid and occultations of Saturn by the Moon.

Occultation in Leo
On 20 March 2014 Leo's brightest star Regulus (bottom right) vanishes briefly from North American skies as the asteroid

The Moon and Saturn
Between March and May 2014 the Moon will pass in front of the ringed planet Saturn no fewer than three times.

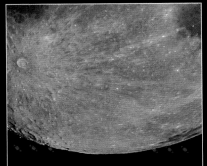

2015

...ear sees two eclipses of
...n, one of which is partial
...e other total (though only
...rctic northern latitudes).
...is also a more widely
...pair of total lunar eclipses,
...spring and one in autumn.

...clipse
...e full Moon passes into Earth's shadow
...eptember, the eclipse will be widely
...n either side of the Atlantic Ocean.

Venus at its best
Our nearest planetary neighbour, Venus, will be
a prominent evening "star" in mid-2015 and an
equally brilliant morning object later in the yea...

| | Full Moon | | New Moon | | Total eclipse of the Moon | | Partial eclipse of the Moon | | Partial eclipse of the Sun | | Annular eclipse of the Sun | | Total eclipse of the Sun |

JANUARY · FEBRUARY · MARCH · APRIL · MAY · JUNE · JULY · AUGUST · SEPTEMBER · OCTOBER · NOVEMBER · DECEMBER

6 June
Venus at greatest
evening elongation

4 September
Mercury at greatest
evening elongation

6 February
Jupiter at opposition,
magnitude -2.6

4 April
Total eclipse of the Moon
visible from western North
America, Australasia, and
east Asia

7 May
Mercury at greatest
evening elongation

14 January
Mercury at greatest
evening elongation

13 September
Partial eclipse of the Sun
visible from southeast Africa
and Antarctica

16 October
Mercury at greatest
morning elongation

24 June
Mercury at greatest
morning elongation

20 March
Total eclipse of the Sun
visible from the Arctic. Partial
eclipse of the Sun visible
from Europe, North Africa,
and northwest Asia

28 September
Total eclipse of the Moon
visible from Europe, Africa,
North America, and
South America

29 December
Mercury at greatest
evening elongation

24 February
Mercury at greatest
morning elongation

23 May
Saturn at opposition,
magnitude 0.0

26 October
Venus at greatest
morning elongation

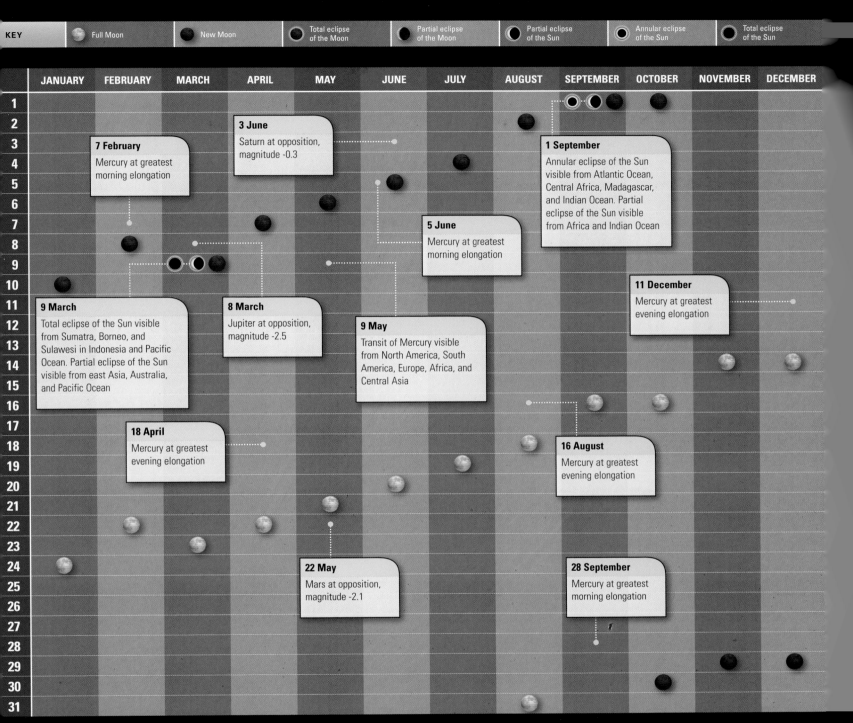

KEY
Full Moon | New Moon | Total eclipse of the Moon | Partial eclipse of the Moon | Partial eclipse of the Sun | Annular eclipse of the Sun | Total eclipse of the Sun

7 February
Mercury at greatest morning elongation

3 June
Saturn at opposition, magnitude -0.3

1 September
Annular eclipse of the Sun visible from Atlantic Ocean, Central Africa, Madagascar, and Indian Ocean. Partial eclipse of the Sun visible from Africa and Indian Ocean

5 June
Mercury at greatest morning elongation

11 December
Mercury at greatest evening elongation

9 March
Total eclipse of the Sun visible from Sumatra, Borneo, and Sulawesi in Indonesia and Pacific Ocean. Partial eclipse of the Sun visible from east Asia, Australia, and Pacific Ocean

8 March
Jupiter at opposition, magnitude -2.5

9 May
Transit of Mercury visible from North America, South America, Europe, Africa, and Central Asia

18 April
Mercury at greatest evening elongation

16 August
Mercury at greatest evening elongation

22 May
Mars at opposition, magnitude -2.1

28 September
Mercury at greatest morning elongation

2016

Alongside two eclipses of the Sun, this year's most interesting astronomical highlight is a transit of Mercury. A similar event will take place in November 2019 – the last transit event until 2032.

Planetary conjunction
The bright planets Jupiter (right) and Saturn (left) are close to each other throughout the year. In January and August Venus joins

Transit of Mercury
On 9 May 2016 the innermost planet Mercury will speed across the face of the Sun in a rare "transit" alignment.

2017

...ear sees both total and
...r eclipses of the Sun, as
...s a widely visible partial lunar
...e. Another interesting celestial
...is an exceptionally close
...ction of Venus and Jupiter
... November.

...lar eclipse
...icular total eclipse will be visible across
... on 21 August 2017. It will last for more
... minutes in many places.

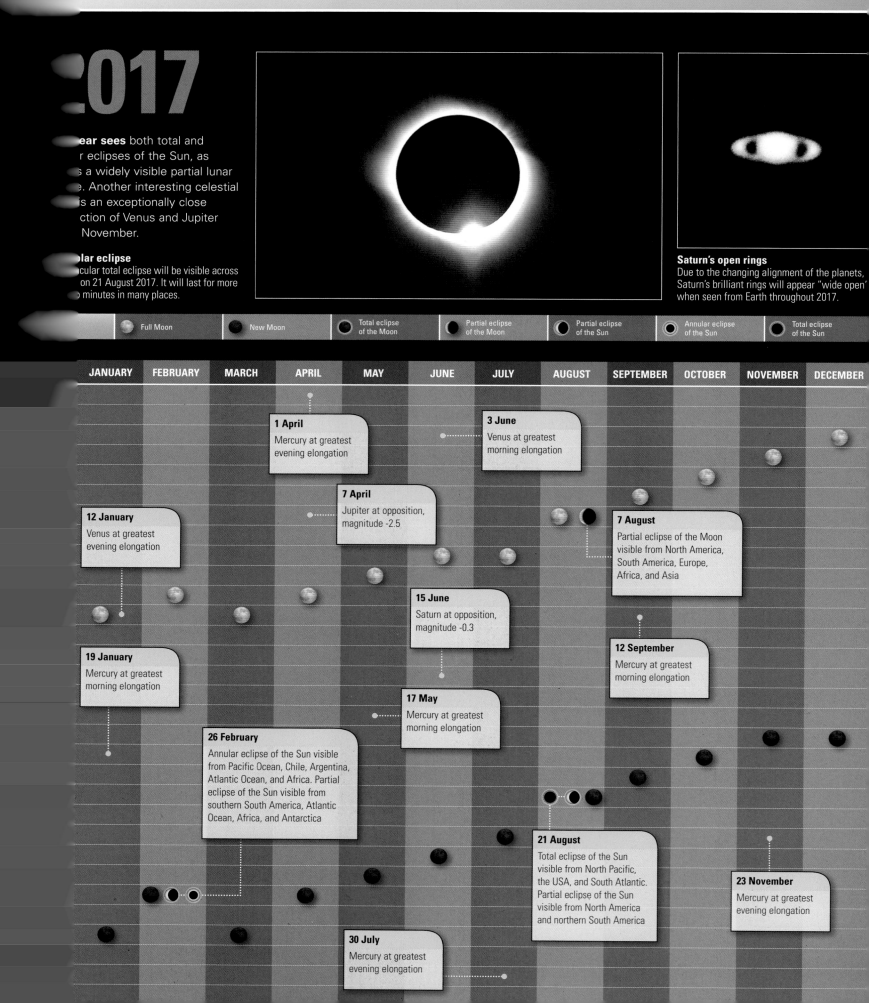

Saturn's open rings
Due to the changing alignment of the planets,
Saturn's brilliant rings will appear "wide open"
when seen from Earth throughout 2017.

| ● Full Moon | ● New Moon | ◐ Total eclipse of the Moon | ◑ Partial eclipse of the Moon | ◐ Partial eclipse of the Sun | ◉ Annular eclipse of the Sun | ◉ Total eclipse of the Sun |

| JANUARY | FEBRUARY | MARCH | APRIL | MAY | JUNE | JULY | AUGUST | SEPTEMBER | OCTOBER | NOVEMBER | DECEMBER |

1 April
Mercury at greatest evening elongation

3 June
Venus at greatest morning elongation

7 April
Jupiter at opposition, magnitude -2.5

7 August
Partial eclipse of the Moon visible from North America, South America, Europe, Africa, and Asia

12 January
Venus at greatest evening elongation

15 June
Saturn at opposition, magnitude -0.3

12 September
Mercury at greatest morning elongation

19 January
Mercury at greatest morning elongation

17 May
Mercury at greatest morning elongation

26 February
Annular eclipse of the Sun visible from Pacific Ocean, Chile, Argentina, Atlantic Ocean, and Africa. Partial eclipse of the Sun visible from southern South America, Atlantic Ocean, Africa, and Antarctica

21 August
Total eclipse of the Sun visible from North Pacific, the USA, and South Atlantic. Partial eclipse of the Sun visible from North America and northern South America

23 November
Mercury at greatest evening elongation

30 July
Mercury at greatest evening elongation

Full Moon ● **New Moon** ◐ **Total eclipse of the Moon** ◐ **Partial eclipse of the Moon** ◐ **Partial eclipse of the Sun** ◉ **Annular eclipse of the Sun** ◐ **Total eclipse of the Sun**

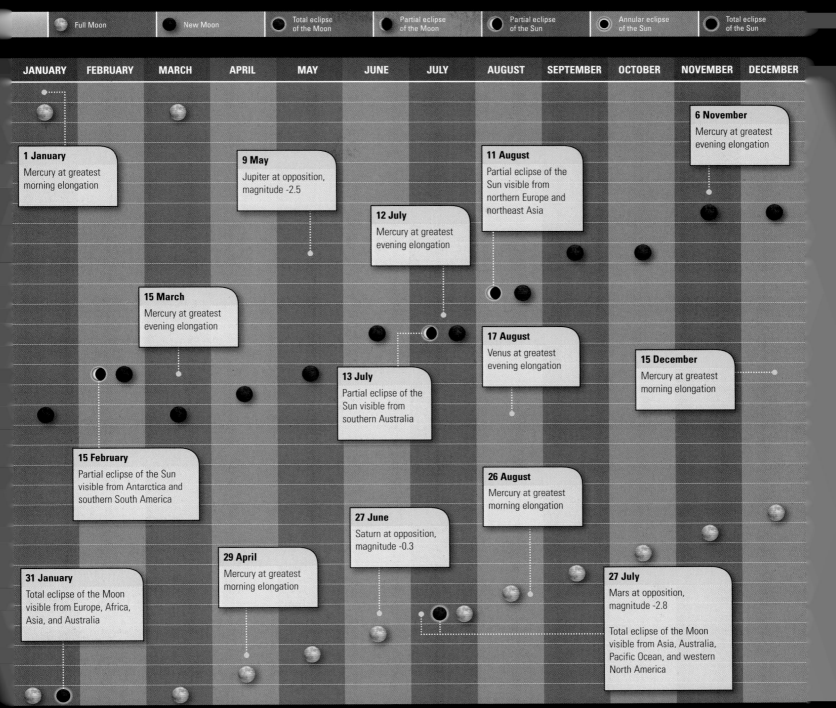

| JANUARY | FEBRUARY | MARCH | APRIL | MAY | JUNE | JULY | AUGUST | SEPTEMBER | OCTOBER | NOVEMBER | DECEMBER |

6 November
Mercury at greatest evening elongation

1 January
Mercury at greatest morning elongation

9 May
Jupiter at opposition, magnitude -2.5

11 August
Partial eclipse of the Sun visible from northern Europe and northeast Asia

12 July
Mercury at greatest evening elongation

15 March
Mercury at greatest evening elongation

17 August
Venus at greatest evening elongation

15 December
Mercury at greatest morning elongation

13 July
Partial eclipse of the Sun visible from southern Australia

15 February
Partial eclipse of the Sun visible from Antarctica and southern South America

26 August
Mercury at greatest morning elongation

27 June
Saturn at opposition, magnitude -0.3

29 April
Mercury at greatest morning elongation

31 January
Total eclipse of the Moon visible from Europe, Africa, Asia, and Australia

27 July
Mars at opposition, magnitude -2.8

Total eclipse of the Moon visible from Asia, Australia, Pacific Ocean, and western North America

2018

of this year's three solar
es are total. However, there
tal lunar eclipse in July, an
tionally close approach of
o Earth around the same time,
close conjunction of Mars and
r in early January.

oons
and March 2018 will both see rare
oons" – events where two full moons
he same calendar month

Close approach of Mars
In July 2018 Mars will come within 58 million km (36 million miles) of Earth, making it exceptionally bright and large in our skies.

2019

ear sees total, annular,
rtial eclipses of the Sun,
l as partial and total eclipses
Moon. Other highlights
e a rare transit of Mercury
unusual disappearing act
iter's satellites.

s great white spot
019 Saturn reaches its northern
ner, when enormous "white spot"
ppear, such as this one seen in 1994.

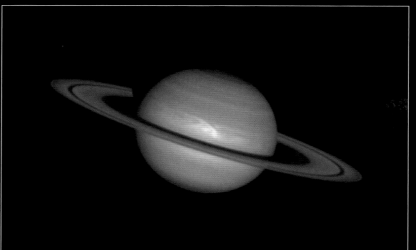

Lonely Jupiter
On 9 November 2019 Jupiter briefly appears
moonless, as all four of its bright satellites are
simultaneously hidden behind the giant planet.

⬤ Full Moon	⬤ New Moon	◐ Total eclipse of the Moon	◑ Partial eclipse of the Moon	◑ Partial eclipse of the Sun	⬤ Annular eclipse of the Sun	◐ Total eclipse of the Sun

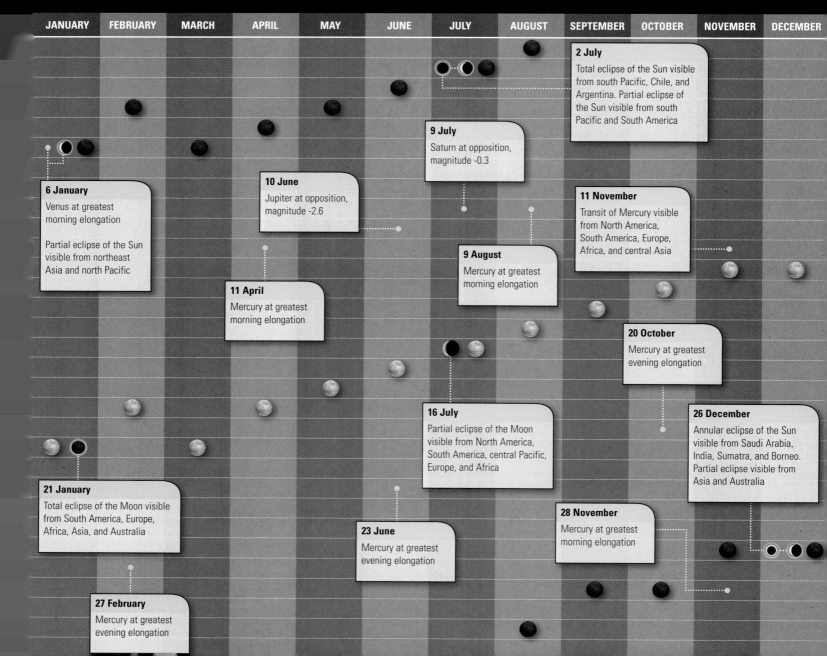

JANUARY	FEBRUARY	MARCH	APRIL	MAY	JUNE	JULY	AUGUST	SEPTEMBER	OCTOBER	NOVEMBER	DECEMBER

2 July
Total eclipse of the Sun visible
from south Pacific, Chile, and
Argentina. Partial eclipse of
the Sun visible from south
Pacific and South America

9 July
Saturn at opposition,
magnitude -0.3

6 January
Venus at greatest
morning elongation

Partial eclipse of the Sun
visible from northeast
Asia and north Pacific

10 June
Jupiter at opposition,
magnitude -2.6

11 November
Transit of Mercury visible
from North America,
South America, Europe,
Africa, and central Asia

9 August
Mercury at greatest
morning elongation

11 April
Mercury at greatest
morning elongation

20 October
Mercury at greatest
evening elongation

16 July
Partial eclipse of the Moon
visible from North America,
South America, central Pacific,
Europe, and Africa

26 December
Annular eclipse of the Sun
visible from Saudi Arabia,
India, Sumatra, and Borneo.
Partial eclipse visible from
Asia and Australia

21 January
Total eclipse of the Moon visible
from South America, Europe,
Africa, Asia, and Australia

23 June
Mercury at greatest
evening elongation

28 November
Mercury at greatest
morning elongation

27 February
Mercury at greatest
evening elongation

GLOSSARY

Aperture The diameter of the main mirror or lens in a telescope or binoculars. A large-aperture telescope can collect more light and detect fainter objects than a small-aperture telescope.

Asterism A recognizable pattern of stars, where the stars are either a part of a constellation or are members of several constellations. An example is the Plough in Ursa Major.

Astrophotography The photography of celestial objects in the night sky, including photography of the Sun and of eclipses.

Binary star Two stars in a mutual orbit around a common centre of mass and bound together gravitationally.

Celestial equator The celestial equivalent of Earth's equator. The celestial equator marks a line where the plane of Earth's equator meets the celestial sphere.

Celestial poles The two points at which the line of Earth's axis, extended outward, meets the celestial sphere and around which the stars appear to revolve.

Celestial sphere The imaginary sphere that surrounds Earth, and upon which all celestial objects appear to lie.

Conjunction An alignment of objects in the night sky, with one passing in front of the other, particularly when a planet lines up with the Sun as viewed from Earth.

Constellation A named area of stars or a designated area of sky around a star pattern. There are currently 88 officially recognized constellations.

Declination A coordinate used in the equatorial coordinate system; it is the celestial equivalent of latitude on Earth. It is measured in degrees above or below the celestial equator,

Deep-sky object Any celestial object lying beyond the Solar System, excluding stars.

Double star Two stars not physically associated with each other, but appearing close together when viewed from Earth.

Eclipse An alignment of a planet or moon with the Sun, which casts a shadow on another celestial body. During a lunar eclipse the Earth's shadow is cast on the Moon, and during a solar eclipse, the Moon's shadow is cast on Earth.

Ecliptic The plane of Earth's orbit around the Sun, or the projection of that plane onto the celestial sphere.

Elliptical galaxy A galaxy that appears as an ellipse. These contain very little gas and dust and are usually devoid of star formations.

Elongation The angular separation between the Sun and an inferior planet as viewed from Earth. Also used at the time of maximum angular separation (greatest elongation) between the inner planets, Mercury or Venus, and the Sun.

Equinox The time when the Sun is vertically overhead Earth's equator, and days and nights are of equal length.

Galaxy A huge mass of stars, gas, and dust linked by gravity. Galaxies may vary from thousands to hundreds of thousands of light-years in range.

Globular cluster A sphere of thousands of stars linked by gravity.

Inferior planets Planets whose orbit around the Sun falls inside Earth's orbit. The two inferior planets are Mercury and Venus.

Light-year The distance travelled by light in one year – 9,460 billion km (5,878 billion miles).

Local group A small cluster of over 30 galaxies; it includes our own galaxy, the Milky Way.

Magnitude The brightness of a celestial object, measured on a numerical scale, where brighter objects are given small or negative numbers, and fainter objects are given larger numbers.

Meteor shower A substantial number of meteors that appear to originate from a common point in the sky.

Multiple stars A system of stars that are bound together gravitationally and are in mutual orbits. Multiple stars have at least three stars and may contain up to a dozen stars.

Nebula A cloud of dust and gas in space, usually made visible by the light of the neighbouring stars.

Open cluster A loose group of up to a few hundred stars, bound by gravity and found in the arms of a galaxy.

Occultation The passage of one celestial body of a larger apparent size over another body of a smaller apparent size, which results in the more distant body being wholly or partially hidden. For example, when the Moon passes over a distant star, it hides the star from our view.

Opposition The time when an outer planet lies on the exact opposite side of Earth from the Sun. The planet is at its closest to Earth and therefore appears brightest at this time.

Orbit The path followed by a planet or other celestial body around the Sun, or by a moon around its parent planet.

Planet A celestial body that has cleared away any planetary debris from its orbit around the Sun and is of a roughly round shape due to its own gravity.

Planetary nebula A glowing shell of dust and gas ejected by a dying star that appears as a luminous planet.

Radiant The point in the sky from which the tracks of meteors that are members of a particular meteor shower appear to originate.

Retrograde motion The rotation of a planet or moon in the opposite direction to its orbit. All of the planets orbit the Sun in the direction of the Sun's rotation: anti-clockwise when viewed from above the Sun's north pole. Most planets also rotate (spin) anti-clockwise. Venus and Uranus have retrograde motion: clockwise compared with their anti-clockwise orbits.

Right ascension (RA) The celestial equivalent of longitude on Earth. It is measured in hours from the position where the ecliptic intersects the celestial equator in spring.

Solar system The family of eight recognized planets and several other celestial bodies such as moons that orbit the Sun.

Spiral galaxy A galaxy that has a distinct set of spiral arms composed of bright young stars. Spiral galaxies are rich in gas and dust, and offer prime conditions for star formation.

Star A huge sphere of glowing plasma that emits heat and light by means of nuclear reactions at its centre.

Superior planets Planets whose orbit around the Sun are outside the orbit of Earth. Mars, Jupiter, Saturn, Uranus, Neptune, and Pluto are the superior planets.

Variable star A star whose magnitude varies over time, brought about by intrinsic or extrinsic changes, such as being eclipsed by another star.

Zodiac A band on either side of the ecliptic, through which the Sun, Moon, and planets appear to travel

INDEX

Note: Months occurring as sub-entries are arranged in month order.

ACKNOWLEDGMENTS

Will Gater
I'd like to thank my family for their continual support, especially Rose, as well as Martha and the team at Dorling Kindersley for all their hard work.

Publisher's acknowledgments
Dorling Kindersley would like to thank the following people for their help in the preparation of this book: Giles Sparrow for editorial and illustration guidance and Almanac text; Paul Drislane for initial design work; additional design help from Fiona McDonald. Indexing Specialists for preparing the index; Lizzie Munsey for proofreading. Additional production help from Sophie Argyris and Luca Frassinetti. DK images: Claire Bowers, Martin Copeland, and Lucy Claxton.

PICTURE CREDITS

The publisher would like to thank the following for their kind permission to reproduce their photographs:

(Key: a-above; b-below/bottom; c-centre; f-far; l-left; r-right; t-top)

2-3 Corbis: Gabe Palmer. **4 Corbis:** Visuals Unlimited. **6-7 Corbis:** Science Faction/Tony Hallas. **9 Corbis:** Myron Jay Dorf (b/Milky Way); NASA/JPL-Caltech (b/Quasar); Science Faction/Tony Hallas (crb); Stocktrek Images (b/Andromeda Galaxy). **European Southern Observatory (ESO):** Digitized Sky Survey 2 (b/Virgo Cluster). **10 Corbis:** Roger Ressmeyer (cra). **13 Corbis:** EPA/Dean Lewins (tr). **Will Gater:** (bl). **14 Corbis:** Gabe Palmer (cla). **15 Corbis:** Frank Lukasseck (bc); Visuals Unlimited (cr). **16-17 Corbis:** Frank Lukasseck. **21 Corbis:** Roger Ressmeyer (cla). **22 Corbis:** Visuals Unlimited (cl) (bl). **23 Robert Gendler:** (br). **Alson Wong:** (cr). **28 NOAO / AURA / NSF:** (cl). **29 Science Photo Library:** Eckhard Slawik (ca). **30 Getty Images:** Visuals Unlimited, Inc./Robert Gendler (bl). **Walter MacDonald:** (cl). **31 Corbis:** Visuals Unlimited (cr). **Science Photo Library:** Celestial Image Co. (br). **36 Corbis:** Roger Ressmeyer (cl). **37 Galaxy Picture Library:** Gordon Garradd (cl). **38 Corbis:** Roger Ressmeyer (bl); Stocktrek Images (cl). **39 European Southern Observatory (ESO):** (br). **NASA and The Hubble Heritage Team (AURA/STScI):** (cr). **44 Getty Images:** David McNew (cl). **45 Yuri Beletsky:** (cl). **46 Corbis:** Stocktrek Images (bl). NOAO / AURA / NSF: (cl). **47 NASA and The Hubble Heritage Team (AURA/STScI):** (br). **Hunter Wilson:** (cr). **52 Corbis:** Roger Ressmeyer (c). **53 Yuri Beletsky:** (cl). **54 NASA and The Hubble Heritage Team (AURA/STScI):** (cl). **NOAO / AURA / NSF:** (bl). **55 NOAO / AURA / NSF:** (cr). **Télescopes à Action Rapide pour les Objets Transitoires:** (br). **61 Corbis:** Amanaimages/Katahira Takashi (cl). **62 Will Gater:** (cl). **NOAO / AURA / NSF:** (bl). **63 Getty Images:** Image Bank/LWA (cr); Visuals Unlimited, Inc./Robert Gendler (br). **68 Will Gater:** (bl). **69 Corbis:** Reuters/Ho (cl). **70 European Southern Observatory (ESO):** Digitized Sky Survey 2 (cl). **NASA:** (bl). **71 Canada-France-Hawaii Telescope:** Jean-Charles Cuillandre (br). **Galaxy Picture Library:** Jeremy Perez (cr). **76 Corbis:** Reuters/Ali Jarekji (bl). **77 Corbis:** Visuals Unlimited (cl). **78 Corbis:** Scott Stulberg (cl). **Getty Images:** Stocktrek Images (bl). **79 Will Gater:** (cr). **NOAO / AURA / NSF:** (br). **85 Corbis:** Stocktrek Images (cl). **86 Frank Barrett:** (bl). **Galaxy Picture Library:** Damian Peach (cl). **87 NASA:** (cr). **NOAO / AURA / NSF:** (br). **93 Alamy Images:** Galaxy Picture Library (cl). **94 Corbis:** Roger Ressmeyer (bl). **Galaxy Picture Library:** Robin Scagell (cl). **95 Anthony Ayiomamitis/perseus.gr:** (cr). **Corbis:** Dennis di Cicco (br). **101 Science Photo Library:** John Chumack (cl). **102 2MASS:** (cl). **NOAO / AURA / NSF:** (bl). **103 Corbis:** Stocktrek Images (br). **NOAO / AURA / NSF:** (cr). **108 Getty Images:** Barcroft Media/Wally Pacholka (cl). **109 Corbis:** Roger Ressmeyer (cl). **110 Corbis:** Roger Ressmeyer (bl). **Galaxy Picture Library:** Damian Peach (cl). **111 Getty Images:** Stocktrek Images (br). **NOAO / AURA / NSF:** (cr). **117 Corbis:** Richard Cummins (br); Reuters/Tim Wimborne (bc). **118 Corbis:** Reuters/Bea Wiharta (tc). **Pete Lawrence:** (tr). **119 Alamy Images:** Galaxy Picture Library (bc). **Getty Images:** SSPL/Jamie Cooper (br). **120 Corbis:** Reuters/Doug Murray (tc). **Jimmy Westlake:** (tr). **121 Science Photo Library:** John Sanford (bc). **Mila Zinkova:** (br). **122 Corbis:** EPA/John Sun (tc); Roger Ressmeyer (tr). **123 Corbis:** Gary Carter (bc); NASA/Bryan Allen (br). **124 Corbis:** (tc). **NASA:** JPL-Caltech (tr).